Springer Theses

Recognizing Outstanding Ph.D. Research

For further volumes:
http://www.springer.com/series/8790

Aims and Scope

The series "Springer Theses" brings together a selection of the very best Ph.D. theses from around the world and across the physical sciences. Nominated and endorsed by two recognized specialists, each published volume has been selected for its scientific excellence and the high impact of its contents for the pertinent field of research. For greater accessibility to non-specialists, the published versions include an extended introduction, as well as a foreword by the student's supervisor explaining the special relevance of the work for the field. As a whole, the series will provide a valuable resource both for newcomers to the research fields described, and for other scientists seeking detailed background information on special questions. Finally, it provides an accredited documentation of the valuable contributions made by today's younger generation of scientists.

Theses are accepted into the series by invited nomination only and must fulfill all of the following criteria

- They must be written in good English.
- The topic should fall within the confines of Chemistry, Physics, Earth Sciences, Engineering and related interdisciplinary fields such as Materials, Nanoscience, Chemical Engineering, Complex Systems and Biophysics.
- The work reported in the thesis must represent a significant scientific advance.
- If the thesis includes previously published material, permission to reproduce this must be gained from the respective copyright holder.
- They must have been examined and passed during the 12 months prior to nomination.
- Each thesis should include a foreword by the supervisor outlining the significance of its content.
- The theses should have a clearly defined structure including an introduction accessible to scientists not expert in that particular field.

Takahiro Sagawa

Thermodynamics of Information Processing in Small Systems

Doctoral Thesis accepted by
The University of Tokyo, Japan

Springer

Author
Dr. Takahiro Sagawa
The Hakubi Center for Advanced Research
Yukawa Institute for Theoretical Physics
Kyoto University
Kyoto
Japan

Supervisor
Prof. Masahito Ueda
Department of Physics
Graduate School of Science
The University of Tokyo
Tokyo
Japan

Chair
Prof. Ryugo Hayano
Department of Physics
Graduate School of Science
The University of Tokyo
Tokyo
Japan

ISSN 2190-5053 ISSN 2190-5061 (electronic)
ISBN 978-4-431-54167-7 ISBN 978-4-431-54168-4 (eBook)
DOI 10.1007/978-4-431-54168-4
Springer Tokyo Heidelberg New York Dordrecht London

Library of Congress Control Number: 2012943971

Printed on acid-free paper

Springer is part of Springer Science+Business Media (www.springer.com)

Parts of this thesis have been published in the following journal articles:

T. Sagawa and M. Ueda, "Second Law of Thermodynamics with Discrete Quantum Feedback Control" Physical Review Letters **100**, 080403(1)–080403(4) (2008).

T. Sagawa and M. Ueda, "Minimum Energy Cost for Thermodynamic Information Processing: Measurement and Information Erasure" Physical Review Letters **102**, 250602(1)–250602(4) (2009).

T. Sagawa and M. Ueda, "Generalized Jarzynski Equality under Nonequilibrium Feedback Control" Physical Review Letters **104**, 090602(1)–090602(4) (2010).

T. Sagawa and M. Ueda, "Sagawa and Ueda Reply" Physical Review Letters **104**, 198904 (2010).

T. Sagawa and M. Ueda, "Erratum: Minimal Energy Cost for Thermodynamic Information Processing: Measurement and Information Erasure" Physical Review Letters **106**, 189901(E) (2011).

T. Sagawa, "Thermodynamics of Information Processing in Small Systems" Progress of Theoretical Physics **127**, 1–56 (2012). DOI: 10.1143/PTP.127.1. Parts of Chaps. 1, 2, 3, 4, 5, 6, 7, and 10 of the thesis are reprinted from this paper with revisions and with the permission. All figures in this paper are reused in the thesis with the permission.

T. Sagawa and M. Ueda, "Nonequilibrium thermodynamics of feedback control" Physical Review E **85**, 021104(1)–021104(16) (2012). Parts of Chaps. 8 and 9 of the thesis are reprinted from this paper with revisions and with the permission. All figures in this paper are reused in the thesis with the permission.

Chairman's Foreword

The origins of the Physics Department of the University of Tokyo may be found in "Bansho- Kurabesho" founded in the Edo era by the Tokugawa Shogunate in 1857. After the Meiji Restoration in 1868, this educational institution was taken over by the Meiji government, and was renamed "Kaisei Gakko". It was in 1875 that a Physics Department opened at Tokyo Kaisei Gakko, which was then integrated into the first university in Japan, the University of Tokyo, founded in 1877. When the University of Tokyo issued its first bachelor degrees in 1879, 12 of the 55 recipients were from the Department of Physics.

Since those early days, the University of Tokyo has grown to acquire a reputation as one of the foremost institutions of tertiary education and scholarship in the world. And the Physics Department's successive generations of faculty and students who engaged in cutting-edge research and pioneering education, have played no small role in this transformation. Of the University's seven Nobel laureates, three are physicists who graduated from the Department, Leona Esaki (awarded 1973), Masatoshi Koshiba (awarded 2002), and Yoichiro Nambu (awarded 2008). The first Fields Medalist from Japan, Kunihiko Kodaira (awarded 1954) was also an alumnus of the Department. Today our staff continue this tradition, engaging in essential research in all aspects of physics, and working to shape the budding next generation of physicists.

It was a great pleasure to learn that our Department of Physics was invited, as one of the top universities in the world for physics, to participate in the Springer Theses initiative. Our first selections are three students among the 51 who wrote doctoral theses in English accepted by our Department in the Fiscal Year 2010. They are Dr. Takahiro Sagawa, Dr. Takayuki Yuasa, and Dr. Takuya Kanazawa. I am happy to express my heartfelt congratulations to these distinguished authors.

Prof. Ryugo Hayano
Chair, The Department of Physics
Graduate School of Science
The University of Tokyo

Supervisor's Foreword

The second law of thermodynamics presupposes a clear-cut distinction between the controllable and uncontrollable degrees of freedom by means of macroscopic operations. The cutting-edge technologies in quantum information and nanoscience seem to require us to abandon such a working hypothesis in favor of the distinction between the accessible and inaccessible degrees of freedom. In his thesis, Takahiro Sagawa discusses the implications of this paradigm shift by addressing the question of how the second law of thermodynamics can be generalized in the presence of a feedback controller, and investigates the minimum work required for measurement and erasure of information.

Ever since the seminal discussion by James Clark Maxwell about an intelligent being presently known as Maxwell's demon, the relationship between thermodynamics and information has attracted much attention because it concerns the foundation of the second law of thermodynamics. In recent years, this relationship has attracted renewed interest because of the advances in nanotechnology which have enabled one to access atomic-scale objects in an controllable manner. Here, Maxwell's demon plays the role of a feedback controller that intervenes between a microscopic object and a macroscopic one. In modern terms, Maxwell's demon is formulated as an information processing device that performs measurement and feedback at the level of thermal fluctuations. By unifying information theory, measurement theory, and the recently developed theory of nonequilibrium statistical mechanics, Takahiro constructs theory of "information thermodynamics," in which information contents and thermodynamic variables are treated on an equal footing. Based on this theory, Takahiro successfully determines the maximum work that can be extracted by the demon and the minimum work that is needed for measurement and information erasure by the demon.

Because information processing is necessarily out of equilibrium, it is an interesting question to ask whether the Jarzynski equality can be generalized in the presence of feedback control and if so how it should be modified. Takahiro has answered this question in the affirmative and generalized several nonequilibrium relations for classical stochastic systems. One of the generalized equalities has recently been verified experimentally by using sub-micron colloidal particles.

Thus, the work presented in this thesis establishes fundamental principles for information processing in small thermodynamic systems, and I expect that the obtained relations can be applied to nanomachines and nanodevices.

Tokyo, March 2012 Masahito Ueda

Acknowledgments

I am most grateful to my supervisor, Prof. Masahito Ueda for fruitful discussions and much valuable advice. I would like to thank the current and previous members of the Ueda group, who have enabled me to enjoy my fruitful research and daily life.

I would like to thank my collaborators Prof. Masaki Sano, Prof. Eiro Muneyuki, Prof. Shoichi Toyabe for an experimental work on Maxwell's demon; Yuji Kurotani and Yu Watanabe for a theoretical work on uncertainty relations and quantum estimation theory; Prof. Sang Wook Kim and Dr. Simone de Liberato for a theoretical work on the quantum Szilard engine.

I would like to thank Prof. Hisao Hayakawa, Prof. Shin-ichi Sasa, Prof. Akira Shimizu, Prof. Yoshiro Takahashi, and Prof. Hal Tasaki for guiding me to non-equilibrium physics, quantum information theory, and quantum optics when I was an undergraduate student.

I would like to thank Prof. Makoto Gonokami, Prof. Mio Murao, Prof. Seiji Miyashita, and Prof. Kazuo Ueda for refereeing my thesis and for valuable discussions.

Moreover, I would like to thank the researchers who have given me the opportunities for a lot of valuable discussions in Japan and overseas.

Finally, I would like to express my gratitude to my family which has greatly supported me.

Contents

Chapter 1
Introduction

Abstract In this thesis, we construct a general theory of thermodynamics of information processing. The background of this research is the recently developed field of nonequilibrium statistical mechanics and quantum and classical information theories. These theories are closely related to the modern technologies used to manipulate and observe small systems; for example, macromolecules and colloidal particles in the classical regime, and quantum-optical systems and quantum dots in the quantum regime. First, we generalize the second law of thermodynamics to the situations in which small thermodynamic systems are subject to quantum feedback control. Second, we generalize the second law of thermodynamics to the measurement and information erasure processes used in the demon's memory. Third, we generalize the nonequilibrium equalities such as the fluctuation theorem and the Jarzynski equality to classical stochastic dynamics in the presence of feedback control. In these results, thermodynamic quantities and information contents are treated on an equal footing. Moreover, the obtained inequalities and equalities are model-independent, so that they can be applied to a broad range of information processing applications. Our findings could be called the second law of "information thermodynamics".

Historically, Maxwell's demon has puzzled numerous physicists over 150 years [1–5]. The demon has shed light on the foundation of thermodynamics and statistical mechanics, because it apparently contradicts the second law of thermodynamics [6–12]. Many researchers have tried to reconcile the demon with the second law.

The first crucial step of the quantitative analysis of the demon was made by Leo Szilard in his paper published in 1929 [13]. He recognized the importance of the concept of information to understand the paradox of Maxwell's demon, about twenty years before an epoch-making paper by Claude E. Shannon [14]. As we will discuss in detail in the next chapter, Szilard considered that, if we take the role of information into account, the demon is shown to be consistent with the second law.

In 1951, Léon Brillouin considered that the key to resolve the paradox of Maxwell's demon is in the measurement process [15]. On the other hand, Charles H. Bennett claimed that the measurement process is irrelevant to resolve the paradox of Maxwell's demon. Instead, in 1982, Bennett argued that the erasure process of

T. Sagawa, *Thermodynamics of Information Processing in Small Systems*,
Springer Theses, DOI: 10.1007/978-4-431-54168-4_1, © Springer Japan 2013

the obtained information is the key to reconcile the demon with the second law [16], based on Landauer's principle proposed by Rolf Landauer in 1961 [17]. The argument by Bennett has been broadly accepted as the resolution of the paradox of Maxwell's demon [2, 3, 15–24]. Recently, however, there have been a lot of advancements [4, 5, 25–33] on the paradox of Maxwell's demon beyond the Landauer-Bennett argument.

We then discuss the modern backgrounds of Maxwell's demon. The recent technologies of controlling small systems have been developed in both classical and quantum regimes. For example, in the classical regime, one can manipulate a single macromolecule or a colloidal particle more precisely than the level of thermal fluctuations at room temperature, by using, for example, optical tweezers. This technique has been applied to investigate biological systems such as the molecular motors [34] (e.g., Kinesins and F_1-ATPases). Moreover, artificial molecular machines [35–39] have also been investigated in both terms of theories and experiments. In the quantum regime, both theories and experiments of quantum measurement and control have been developed at the level of a single atom or a single photon.

In particular with these developments, powerful theories have been established in nonequilibrium statistical mechanics and quantum information theory.

In nonequilibrium statistical mechanics, thermodynamic aspects of small systems have become more and more important [40–42]. In the classical regime, macromolecules and colloidal particles are typical examples of the small thermodynamic systems. In the quantum regime, quantum dots can be regarded as a typical example. The crucial feature of such small thermodynamic systems is that their dynamics is stochastic; their thermal or quantum fluctuations become the same order of magnitude as the averages of the physical quantities. Therefore, the fluctuations play crucial roles to understand the dynamics of such systems. In this thesis, we will mainly focus on the thermal fluctuations in terms of nonequilibrium statistical mechanics. Since 1993, a lot of equalities that are universally valid in nonequilibrium stochastic systems have been found [43–50]. A prominent result is the fluctuation theorem that enables us to quantitatively understand the probability of the stochastic violation of the second law of thermodynamics in small systems. Another prominent result is the Jarzynski equality that expresses the second law of thermodynamics by an equality rather than an inequality. From the first cumulant of the Jarzynski equality, we can reproduces the conventional second law of thermodynamics that is expressed in terms of an inequality. The second law of thermodynamics can be shown to still hold on average even in small systems without Maxwell's demon, while the second law is stochastically violated in small systems due to the thermal fluctuations.

On the other hand, quantum measurement theory has been established [51–55], and has been applied to a lot of systems including quantum-optical systems [56–58]. The concepts of positive operator-valued measures (POVMs) and Kraus operators (measurement operators) play crucial roles, which enable us quantitatively calculate the probability distributions of the outcomes and the backactions of quantum measurements. These theoretical concepts correspond to a lot of experimental situations in which one performs indirect measurements by using a measurement apparatus. Moreover, based on the quantum measurement theory, quantum information theory [55] has also been developed, which is a generalization of classical information theory proposed by Shannon [14, 59].

Based on these backgrounds, Maxwell's demon and thermodynamics of information processing have been attracted renewed attentions [60–98]. In particular, Maxwell's demon can be characterized as a feedback controller acting on thermodynamic systems [33, 63, 75]. Here, "feedback" means that a control protocol depends on measurement outcomes obtained from the controlled system [99, 100]. Feedback control is useful to experimentally realize intended dynamical properties of small systems both in classical and quantum systems. While feedback control has a long history more than 50 years in science and engineering, the modern technologies enable us to control the thermal fluctuation at the level of $k_B T$ with k_B being the Boltzmann constant and T being the temperature. In fact, recently, Szilard-type Maxwell's demon has been realized for the first time [84], by using a real-time feedback control of a colloidal particle on a ratchet-type potential.

In this thesis, we develop the general theory of thermodynamics of information processing in small systems, by using both nonequilibrium statistical mechanics and quantum information theory. The significances of this study are:

- It sheds new light on the foundations of thermodynamics and statistical mechanics.
- It is applicable to the analysis of the thermodynamic properties of a broad class of information processing.

In particular, we generalize the second law of thermodynamics to information processing such as the feedback control, the measurement, and the information erasure. The generalized second laws involve the terms of information contents, and identify the fundamental lower bounds of the energy costs that are needed for the information processing in both classical and quantum regimes.

We also propose an explicit counter-example against Bennett's argument to resolve the paradox of Maxwell's demon, and propose a general and quantitative way to reconcile the demon with the second law. The paradox of Maxwell's demon has been essentially resolved by this argument.

Moreover, we will generalize the nonequilibrium equalities in the presence of feedback control. These generalized equalities reproduce the generalized second law with feedback control as the properties of their first cumulant. The generalized Jarzynski equality, which is one of the generalized equalities, has been experimentally verified with a Szilard-type Maxwell's demon [84].

This thesis is organized as follows. The main results of this thesis are in Chaps. 6, 7, and 9.

In Chap. 2, we review the basic concepts and the history of the problem of Maxwell's demon. Starting from the review of the original gedankenexperiment by Maxwell, we discuss the arguments by Szilard, Brillouin, Landauer, and Bennett.

To generally formulate Maxwell's demon in a model-independent way, we need modern information theory which will be reviewed in Chaps. 3 and 4. In Chap. 3, we focus on the classical aspects: we review the general formulations of classical stochastic dynamics, information, and measurement. The key concepts in this chapter are the Shannon information and the mutual information. In Chap. 4, we focus on the quantum aspects of information theory. Starting from the formulation of the dynamics of unitary and nonunitary quantum systems, we shortly review quantum measurement theory and quantum information theory. In particular, we introduce the concept of QC-mutual information and prove its important properties. Moreover, we discuss that the quantum formulation includes the classical one as a special case, by pointing out the quantum-classical correspondence. Therefore, while we discuss only quantum formulations in Chaps. 5, 6, and 7, the formulations and results includes classical ones.

In Chap. 5, we review the possible derivations of the second law of thermodynamics based on statistical mechanics. In particular, we discuss the proof of the second law in terms of the unitary evolution of the total system including multi-heat baths with the initial canonical distributions. This approach to prove the second law is the standard one in modern nonequilibrium statistical mechanics. We derive several inequalities including Kelvin's inequality, the Clausius inequality, and its generalization to nonequilibrium initial and final states. The proof is independent of the size of the thermodynamic system, and can be applied to small thermodynamic systems.

Chapter 6 is the first main part of this thesis. We derive a generalized second law of thermodynamics with a single quantum measurement and quantum feedback control [33, 75], by involving the measurement and feedback to the proof in Chap. 5. The QC-mutual information introduced in Chap. 4 characterizes the upper bound of the additional work that can be extracted from heat engines with the assistance of feedback control, or Maxwell's demon.

In Chap. 7, we discuss the thermodynamic aspects of Maxwell's demon itself (or the memory of the feedback controller), which is the second main part of this thesis. Starting from the formulation of the memory that stores measurement outcomes, we identify the lower bound of the thermodynamic energy costs for the measurement and the information erasure [31, 33]. Our result for the erasure includes Landauer's principle as a special case.

By using the general results in Chaps. 6 and 7, we can essentially reconcile Maxwell's demon with the second law of thermodynamics, which leads to a novel and general physical picture of the resolution of the paradox of Maxwell's demon. Moreover, our results in Chaps. 6 and 7 can be regarded as the generalizations of the second law of thermodynamics to information processing processes: feedback, measurement, and information erasure.

In Chap. 8, we review the stochastic aspects of classical thermodynamic systems with multiple heat baths. In particular, we formulate and derive the detailed fluctuation theorem, which is one of the most fundamental results in modern nonequilibrium physics. We discuss that the detailed fluctuation theorem leads to several important nonequilibrium relations such as the Jarzynski equality. As a specific example, we review the Markovian dynamics that satisfies the detailed balance.

Chapter 9 is the third main part of this thesis. We generalize the nonequilibrium relations such as the detailed fluctuation theorem and the Jarzynski equality to the situations that classical thermodynamic systems are subject to feedback control with multiple measurements [80, 98]. We discuss the two types of generalizations:

one involves the term of the mutual information obtained by the measurements, and the other involves the term that characterizes the efficacy of feedback control.

In Chap. 10, we conclude this thesis.

References

1. J.C. Maxwell, *Theory of Heat* (Appleton, London, 1871)
2. H.S. Leff, A.F. Rex (eds.), *Maxwell's Demon 2: Entropy, Classical and Quantum Information, Computing* (Princeton University Press, New Jersey, 2003)
3. K. Maruyama, F. Nori, V. Vedral, Rev. Mod. Phys. **81**, 1 (2009)
4. O.J.E. Maroney, Information processing and thermodynamic entropy, in The Stanford Encyclopedia of Philosophy, ed. by E.N. Zalta (Fall 2009 Edition).
5. T. Sagawa, M. Ueda, Information thermodynamics: Maxwell's demon in nonequilibrium dynamics, in Nonequilibrium Statistical Physics of Small Systems: Fluctuation Relations and Beyond, ed. by R. Klages, W. Just, C. Jarzynski (Wiley-VCH, Weinheim, 2012), arXiv:1111.5769 (2011) (To appear).
6. S. Carnot, *Réflexions sur la pussance motrice du feu et sur les machines propresà développer atte puissance* (Bachelier, Paris, 1824)
7. L. Tisza, P.M. Quay, Ann. Phys. **25**, 48 (1963)
8. E.H. Lieb, J. Yngvason, Phys. Rept. **310**, 1 (1999)
9. H.B. Callen, *Thermodynamics and an Introduction to Thermostatistics*, 2nd edn. (Wiley, New York, 1985)
10. H. Tasaki, *Thermodynanmics–From a Modern Point of View* (Baifu-kan, Chiyoda, 2000). (in Japanese)
11. S. Sasa, *Introduction to Thermodynamics* (Kyoritsu, Tokyo, 2000). (in Japanese)
12. A. Shimizu, *Principles of Thermodynamics* (University of Tokyo Press, Tokyo, 2007). (in Japanese)
13. L. Szilard, Z. Phys. **53**, 840 (1929)
14. C. Shannon, Bell Syst. Tech. J. **27**(371–423), 623–656 (1948)
15. L. Brillouin, J. Appl. Phys. **22**, 334 (1951)
16. C.H. Bennett, Int. J. Theor. Phys. **21**, 905 (1982)
17. R. Landauer, IBM J. Res. Dev. **5**, 183 (1961)
18. W.H. Zurek, Nature **341**, 119 (1989)
19. W.H. Zurek, Phys. Rev. A **40**, 4731 (1989)
20. K. Shizume, Phys. Rev. E **52**, 3495 (1995)
21. R. Landauer, Science **272**, 1914 (1996)
22. H. Matsueda, E. Goto, K-F. Loe, RIMS Kôkyûroku **1013**, 187 (1997)
23. B. Piechocinska, Phys. Rev. A **61**, 062314 (2000)
24. C.H. Bennett, Stud. Hist. Philos. Mod. Phys. **34**, 501 (2003)
25. A.E. Allahverdyan, T.M. Nieuwenhuizen, Phys. Rev. E **64**, 0561171 (2001)
26. C. Horhammer, H. Buttner, J. Stat. Phys. **133**, 1161 (2008)
27. M.M. Barkeshli, arXiv:cond-mat/0504323 (2005).
28. J.D. Norton, Stud. Hist. Philos. Mod. Phys. **36**, 375 (2005)
29. O.J.E. Maroney, Phys. Rev. E **79**, 031105 (2009)
30. S. Turgut, Phys. Rev. E **79**, 041102 (2009)
31. T. Sagawa, M. Ueda, Phys. Rev. Lett. **102**, 250602 (2009)
32. T. Sagawa, M. Ueda, Phys. Rev. Lett. 106, 189901(E) (2011).
33. T. Sagawa, Prog. Theor. Phys. **127**, 1 (2012)
34. M. Schliwa, G. Woehlke, Nature **422**, 751–765 (2003)
35. Y. Shirai et al., Nano Lett. **5**, 2330 (2005)
36. V. Serreli et al., Nature **445**, 523 (2007)

37. S. Rahav, J. Horowitz, C. Jarzynski, Phys. Rev. Lett. **101**, 140602 (2008)
38. E.R. Kay, D.A. Leigh, F. Zerbetto, Angew. Chem. **46**, 72 (2007)
39. H. Gu et al., Nature **465**, 202 (2010)
40. K. Sekimoto, Prog. Theor. Phys. Suppl. **130**, 17 (1998)
41. C. Bustamante, J. Liphardt, F. Ritort, Phys. Today **58**, 43 (2005)
42. U. Seifert, Eur. Phys. J. B **64**, 423 (2008)
43. D.J. Evans, E.G.D. Cohen, G.P. Morriss, Phys. Rev. Lett. **71**, 2401 (1993)
44. G. Gallavotti, E.G.D. Cohen, Phys. Rev. Lett. **74**, 2694 (1995)
45. C. Jarzynski, Phys. Rev. Lett. **78**, 2690 (1997)
46. J. Kurchan, J. Phys. A: Math. Gen. **31**, 3719 (1998)
47. G.E. Crooks, J. Stat. Phys. **90**, 1481 (1998)
48. G.E. Crooks, Phys. Rev. E **60**, 2721 (1999)
49. C. Maes, J. Stat. Phys. **95**, 367 (1999)
50. C. Jarzynski, J. Stat. Phys. **98**, 77 (2000)
51. J. von Neumann, *Mathematische Grundlagen der Quantumechanik* (Springer, Berlin, 1932). [Eng. trans. R.T. Beyer, Mathematical Foundations of Quantum Mechanics (Prinston University Press, Princeton, 1955)]
52. E.B. Davies, J.T. Lewis, Commun. Math. Phys. **17**, 239 (1970)
53. K. Kraus, Ann. Phys. **64**, 311 (1971)
54. M. Ozawa, J. Math. Phys. **25**, 79 (1984)
55. M.A. Nielsen, I.L. Chuang, *Quantum Computation and Quantum Information* (Cambridge University Press, Cambridge, 2000)
56. H.P. Breuer, F. Petruccione, *The Theory of Open Quantum Systems* (Oxford University Press, Oxford, 2002)
57. K. Koshino, A. Shimizu, Phys. Rept. **412**, 191 (2005)
58. H.M. Wiseman, G.J. Milburn, *Quantum Measurement and Control* (Cambridge University Press, Cambridge, UK, 2010)
59. T.M. Cover, J.A. Thomas, *Elements of Information Theory* (Wiley, New York, 1991)
60. S. Lloyd, Phys. Rev. A **39**, 5378 (1989)
61. C.M. Caves, Phys. Rev. Lett. **64**, 2111 (1990)
62. S. Lloyd, Phys. Rev. A **56**, 3374 (1997)
63. M.A. Nielsen, C.M. Caves, B. Schumacher, H. Barnum, Proc. R. Soc. Lond. A **454**, 277 (1998)
64. H. Touchette, S. Lloyd, Phys. Rev. Lett. **84**, 1156 (2000)
65. W.H. Zurek, arXiv:quant-ph/0301076 (2003).
66. T.D. Kieu, Phys. Rev. Lett. **93**, 140403 (2004)
67. A.E. Allahverdyan, R. Balian, ThM Nieuwenhuizen, J. Mod. Opt. **51**, 2703 (2004)
68. H. Touchette, S. Lloyd, Phys. A **331**, 140 (2004)
69. H.T. Quan, Y.D. Wang, Y-x. Liu, C. P. Sun, F. Nori. Phys. Rev. Lett. **97**, 180402 (2006)
70. F.J. Cao, L. Dinis, J.M.R. Parrondo, Phys. Rev. Lett. **93**, 040603 (2004)
71. K.H. Kim, H. Qian, Phys. Rev. Lett. **93**, 120602 (2004)
72. K.H. Kim, H. Qian, Phys. Rev. E **75**, 022102 (2007)
73. B.J. Lopez, N.J. Kuwada, E.M. Craig, B.R. Long, H. Linke, Phys. Rev. Lett. **101**, 220601 (2008)
74. A.E. Allahverdyan, D.B. Saakian, Europhys. Lett. **81**, 30003 (2008)
75. T. Sagawa, M. Ueda, Phys. Rev. Lett. **100**, 080403 (2008)
76. K. Jacobs, Phys. Rev. A **80**, 012322 (2009)
77. F.J. Cao, M. Feito, Phys. Rev. E **79**, 041118 (2009)
78. M. Feito, J.P. Baltanas, F.J. Cao, Phys. Rev. E **80**, 031128 (2009)
79. F.J. Cao, M. Feito, H. Touchette, Phys. A **388**, 113 (2009)
80. T. Sagawa, M. Ueda, Phys. Rev. Lett. **104**, 090602 (2010)
81. M. Ponmurugan, Phys. Rev. E **82**, 031129 (2010)
82. Y. Fujitani, H. Suzuki, J. Phys. Soc. Jpn. **79**, 104003 (2010)
83. J.M. Horowitz, S. Vaikuntanathan, Phys. Rev. E **82**, 061120 (2010)

84. S. Toyabe, T. Sagawa, M. Ueda, E. Muneyuki, M. Sano, Nat. Phys. **6**, 988 (2010)
85. S.W. Kim, T. Sagawa, S. De Liberato, M. Ueda, Phys. Rev. Lett. **106**, 070401 (2011)
86. Y. Morikuni, H. Tasaki, J. Stat. Phys. **143**, 1 (2011)
87. S. Ito, M. Sano, Phys. Rev. E **84**, 021123 (2011)
88. J.M. Horowitz, J.M.R. Parrondo, Europhys. Lett. **95**, 10005 (2011)
89. D. Abreu, U. Seifert, Europhys. Lett. **94**, 10001 (2011)
90. S. Vaikuntanathan, C. Jarzynski, Phys. Rev. E **83**, 061120 (2011)
91. T. Sagawa, J. Phys. Conf. Ser. **297**, 012015 (2011)
92. H. Dong, D.Z. Xu, C.Y. Cai, C.P. Sun, Phys. Rev. E **83**, 061108 (2011)
93. D.V. Averin, M. Möttönen, J.P. Pekola, Phys. Rev. B **84**, 245448 (2011)
94. J.M. Horowitz, J.M.R. Parrondo, New J. Phys. **13**, 123019 (2011)
95. L. Granger, H. Kantz, Phys. Rev. E **84**, 061110 (2011)
96. S. Lahiri, S. Rana, A.M. Jayannavar, J. Phys. A: Math. Theor. **45**, 065002 (2012)
97. Y. Lu, G.L. Long, Phys. Rev. E **85**, 011125 (2012)
98. T. Sagawa, M. Ueda, Phys. Rev. E **85**, 021104 (2012)
99. J.C. Doyle, B.A. Francis, A.R. Tannenbaum, *Feedback Control Theory* (Macmillan, New York, 1992)
100. K.J. Åstrom, R.M. Murray, *Feedback Systems: An Introduction for Scientists and Engineers* (Princeton University Press, Princeton, 2008)

Chapter 2
Review of Maxwell's Demon

Abstract The paradox of Maxwell's demon was proposed in a letter from James C. Maxwell to Peter G. Tait for the first time. In the letter, Maxwell mentioned his gedankenexperiment of "a being whose faculties are so sharpened that he can follow every molecule" [1]. The being may be like a tiny fairy, and may violate the second law of thermodynamics. In 1874, William Thomson, who is also well-known as Lord Kelvin, gave it an impressive but opprobrious name—"demon." Later, Leo Szilard proposed an important model of the demon, which quantitatively connects the thermodynamic work to information [2]. Since then, numerous researchers have been discussed the foundation of the second law of thermodynamics in terms of Maxwell's demon [3–16]. In this chapter, we review the historical arguments and the basic ideas related to the problem of the demon. The modern aspects of the demon [5, 6, 17–25] will be discussed in the following chapters.

2.1 Original Maxwell's Demon

First of all, we consider the original version of the demon proposed by Maxwell (see also Fig. 2.1) [1]. A classical ideal gas is in a box that is adiabatically separated from the environment. In the initial state, the gas is in thermal equilibrium at temperature T. Suppose that a barrier is inserted at the center of the box, and a small door is attached to the barrier. A small being, which is named as a "demon" by Kelvin, is in the front of the door. It has the capability of measuring the velocity of each molecule in the gas, and it opens or closes the door depending on the measurement outcomes. If a molecule whose velocity is higher than the averaged one comes from the left box, then the demon opens the door. If a molecule whose velocity is slower than the average one comes from the right box, then the demon also opens the door. Otherwise the door is closed. By repeating this operation again and again, the gas in the left box gradually becomes cooler than the initial temperature, and the gas in the right box becomes hotter. After all, the demon is able to adiabatically create the temperature

T. Sagawa, *Thermodynamics of Information Processing in Small Systems*, Springer Theses, DOI: 10.1007/978-4-431-54168-4_2, © Springer Japan 2013

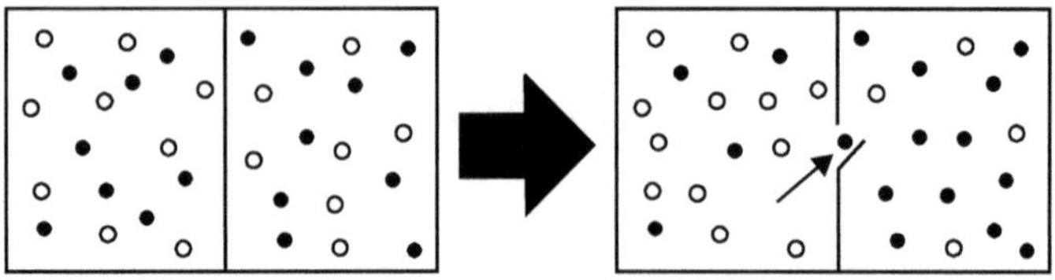

Fig. 2.1 The original gedankenexperiment of Maxwell's demon (reproduced from Ref. [25] with permission). A *white* (*black*) *particle* indicates a molecule whose velocity is slower (faster) than the average. The demon adiabatically realizes a temperature difference by measuring the velocities of molecules and controlling the door based on the measurement outcomes

difference starting from the initial uniform temperature. In other words, the entropy of the gas is more and more decreased by the action of the demon, though the box is adiabatically separated from the outside. This apparent contradiction to the second law has been known as the paradox of Maxwell's demon.

The important point of this gedankenexperiment is that the demon can perform the measurement at the single-molecule level, and can control the door based on the measurement outcomes (i.e., the molecule's velocity is faster or slower than the average), which implies the demon can perform feedback control of the thermal fluctuation.

2.2 Szilard Engine

The first crucial model of Maxwell's demon that quantitatively clarified the role of the information was proposed by Szilard in 1929 [2]. The setup by Szilard seems to be a little different from the Maxwell's one, but the essence—the role of the measurement and feedback—is the same.

Let us consider a classical single molecule gas in an isothermal box that contacts with a single heat bath at temperature T. The Szilard's engine consists of the following five steps (see also Fig. 2.2).

Step 1: *Initial state*. In the initial state, a single molecule is in thermal equilibrium at temperature T.

Step 2: *Insertion of the barrier*. We next insert a barrier at the center of the box, so that we divide the box into two boxes. In this stage, we do not know which box the molecule is in. In the ideal case, we do not need any work for this insertion process.

Step 3: *Measurement*. The demon then measures the position of the molecule, and finds whether the molecule is in "left" or "right." This measurement is assumed to be error-free. The information obtained by the demon is 1 bit, which equals ln 2 nat in the natural logarithm, corresponding to the binary outcome of "left" or "right." The rigorous formulation of the concept of information will be discussed in the next chapter.

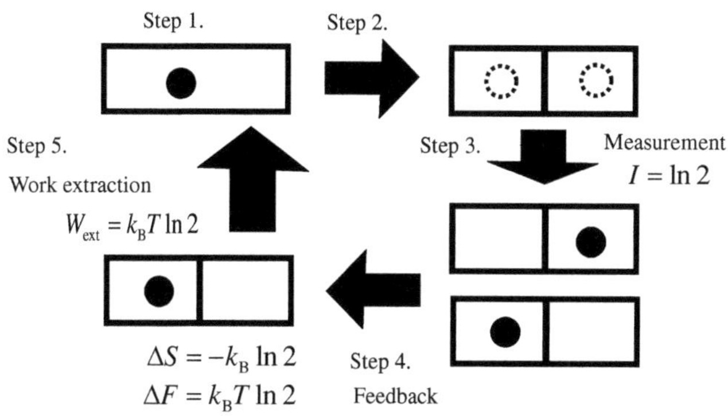

Fig. 2.2 Schematic of the Szilard engine (reproduced from Ref. [25] with permission). *Step 1*: Initial equilibrium state of a single molecule at temperature T. *Step 2*: Insertion of the barrier. *Step 3*: Measurement of the position of the molecule. The demon gets $I = \ln 2$ nat of information. *Step 4*: Feedback control. The demon moves the box to the left only if the measurement outcome is "right." *Step 5*: Work extraction by the isothermal and quasi-static expansion. The state of the engine then returns to the initial one. During this isothermal cycle, we can extract $k_B T \ln 2$ of work

Step 4: *Feedback*. The demon next performs the control depending on the measurement outcome, which is regarded as a feedback control. If the outcome is "left," then the demon does nothing. On the other hand, if the outcome is "right," then the demon quasi-statically moves the right box to the left position. No work is needed for this process, because the motion of the box is quasi-static. After this feedback process, the state of the system is independent of the measurement outcome; the post-feedback state is always "left."

Step 5: *Extraction of the work*. We then expand the left box quasi-statically and isothermally, so that the system returns to the initial state. Since the expansion is quasi-static and isothermal, the equation of states of the single-molecular ideal gas always holds:

$$pV = k_B T, \tag{2.1}$$

where p is the pressure, V is the volume, and k_B is the Boltzmann constant. Therefore, we extract $W_{ext} = k_B T \ln 2$ of work during this expansion, which is followed from

$$W_{ext} = \int_{V_0/2}^{V_0} dV \frac{k_B T}{V} = k_B T \ln 2, \tag{2.2}$$

where V_0 is the initial volume of the box.

During the total process described above, we can extract the positive work of $k_B T \ln 2$ from the isothermal cycle with the assistance of the demon. This apparently contradicts the second law of thermodynamics for isothermal processes known as Kelvin's principle, which states that we cannot extract any positive work from any isothermal cycle in the presence of a single heat bath. In fact, if one could violate Kelvin's principle, one was able to create a perpetual motion of the second kind. Therefore, the fundamental problem is the following:

- Is the Szilard engine a perpetual motion of the second kind?
- If not, what compensates for the excess work of $k_B T \ln 2$?

This is the problem of Maxwell's demon.

The crucial feature of the Szilard engine lies in the fact that the extracted work of $k_B T \ln 2$ is proportional to the obtained information $\ln 2$ with the coefficient of $k_B T$. Therefore, it would be expected that the information plays a key role to resolve the paradox of Maxwell's demon. In fact, from Step 2 to Step 4, the demon decreases $k_B \ln 2$ of physical entropy corresponding to the thermal fluctuation between "left" or "right," by using $\ln 2$ of information. Immediately after the measurement in Step 3, the state of the molecule and the measurement outcome are perfectly correlated, which implies that the demon has the perfect information about the measured state (i.e., "left" or "right"). However, immediately after the feedback in Step 5, the state of the molecule and the measurement outcome is no longer correlated. Therefore, we can conclude that the demon uses the obtained information as a resource to decrease the physical entropy of the system. This is the bare essential of the Szilard engine. On the other hand, the decrease of $k_B \ln 2$ of the entropy means the increase of $k_B T \ln 2$ of the Helmholtz free energy, because $F = E - TS$ holds with F being the free energy, E being the internal energy, and S being the entropy. Therefore, the free energy is increased by $k_B T \ln 2$ during the feedback control by the demon, and the increase in the free energy has been extracted as the work in Step 5. This is how the information has been used in the Szilard engine to extract the positive work.

Szilard pointed out that the increase of the entropy in the memory of the demon compensates for the decrease of the entropy of $k_B \ln 2$ by feedback control. In fact, the memory of the demon, which stores the obtained information of "left" or "right," is itself a physical system, and the fluctuation of the measurement outcome implies an increase in the physical entropy of the memory. In fact, to decrease $k_B \ln 2$ of the physical entropy of the controlled system (i.e., the Szilard engine), at least the same amount of physical entropy must increases elsewhere corresponding to the obtained information, so that the second law of thermodynamics for the total system of the Szilard engine and demon's memory is not violated. This is a crucial observation made by Szilard. However, it was not yet so clear which process actually compensates for the excess work of $k_B T \ln 2$. This problem has been investigated by Brillouin, Landauer, and Bennett.

2.3 Brillouin's Argument

In 1951, Brillouin made an important argument on the problem of Maxwell's demon [7]. He considered that the excess work of $k_B T \ln 2$ is compensated for by the work that is needed for the measurement process by the demon.

He considered that the demon needs to shed a probe light, which is at least a single photon, to the molecule to detect its position. However, if the temperature of the heat bath is T, there must be the background radiation around the molecule. The energy of a photon of the background radiation is about $k_B T$. Therefore, to distinguish the probe photon from the background photons, the energy of the probe photon should be much greater than that of the background photons:

$$\hbar\omega_P \gg k_B T, \tag{2.3}$$

where ω_P is the frequency of the probe photon. Inequality (2.3) may imply

$$W_{\text{meas}} = \hbar\omega_P > k_B T \ln 2, \tag{2.4}$$

which means that the energy cost W_{meas} that is needed for the measurement should be larger than the excess work of $k_B T \ln 2$. Therefore, Brillouin considered that the energy cost for the measurement process compensates for the excess work, so that we cannot extract any positive work from the Szilard engine.

We note that, from the modern point of view, Brillouin's argument depends on a specific model to measure the position of the molecule.

2.4 Landauer's Principle

On the other hand, in his paper published in 1961 [8], Landauer considered the fundamental energy cost that is needed for the erasure of the obtained information from the memory. He propose an important observation, which is known as Landauer's principle today: to erase one bit (= ln 2 nat) of information from the memory in the presence of a single heat bath at temperature T, at least $k_B T \ln 2$ of heat should be dissipated from the memory to the environment.

This statement can be understood as follows. Before the information erasure, the memory stores ln 2 of information, which can be represented by "0" and "1." For example, as shown in Fig. 2.3, if the particle is in the left well, the memory stores the information of "0," while if the particle is in the right well, the memory stores information of "1." This information storage corresponds to $k_B \ln 2$ of entropy of the memory. After the information erasure, the state of the memory is reset to the standard state, say "0," with unit probability as shown in Fig. 2.3. The entropy of the memory then decreases by $k_B \ln 2$ during the information erasure. According to the conventional second law of thermodynamics, the decrease of the entropy in any isothermal process should be accompanied by the heat dissipation to the environment. Therefore, during the erasure process, at least $k_B T \ln 2$ of heat is dissipated from the memory to the heat bath, corresponding to the decrease of the entropy of $k_B \ln 2$. This is the physical origin of Landauer's principle, which is closely related to the second law of thermodynamics.

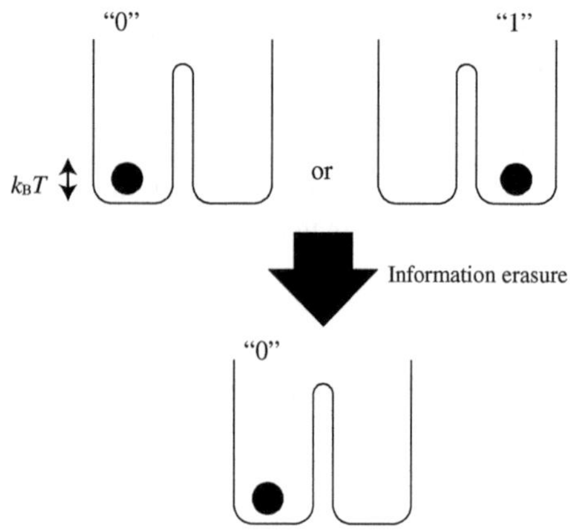

Fig. 2.3 Schematic of information erasure (reproduced from Ref. [25] with permission). Before the erasure, the memory stores information "0" or "1." After the erasure, the memory goes back to the standard state "0" with unit probability

If the internal energies of "0" and "1" are degenerate, we need the same amount of the work as the heat to compensates for the heat dissipation. Therefore, Landauer's principle can be also stated as

$$W_{\rm eras} \geq k_{\rm B} T \ln 2, \tag{2.5}$$

where $W_{\rm eras}$ is the work that is needed for the erasure process.

The argument by Landauer seems to be very general and model-independent, because it is a consequence of the second law of thermodynamics. However, the proof of Landauer's principle based on statistical mechanics has been given only for a special type of memories that is represented by the symmetric binary potential described in Fig. 2.3 [12, 15]. We note that Goto and his collaborators argued that there is a counter-example of Landauer's principle [14].

2.5 Bennett's Argument

In 1982, Bennett proposed an explicit example in which we do not need any energy cost to perform a measurement, which implies that there is a counter-example against Brillouin's argument [9]. Moreover, Bennett argued that, based on Landauer's principle (2.5), we always need the energy cost for information erasure from demon's memory, which compensates for the excess work of $k_{\rm B} T \ln 2$ that is extracted from the Szilard engine by the demon.

His proposal of the resolution of the paradox of Maxwell's demon can be summarized as follows. To make the total system of the Szilard engine and demon's memory a thermodynamic cycle, we need to reset the memory's state which corresponds to information erasure. While we do not necessarily need for the work for the measurement, at least $k_{\rm B} T \ln 2$ of work is always needed the work for the erasure.

Fig. 2.4 Logical reversibility and irreversibility (reproduced from Ref. [25] with permission). **a** Logically reversible measurement process. **b** Logically irreversible erasure process

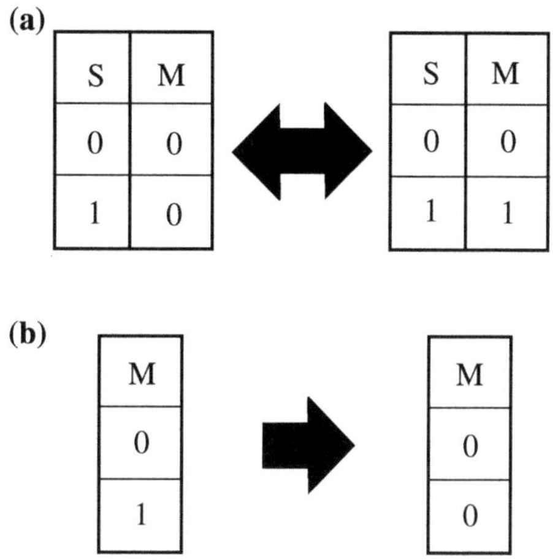

Therefore, the information erasure is the key to reconcile the demon with the second law of thermodynamics.

Bennett's argument is also related to the concept of logical reversibility of classical information processing. For example, the classical measurement process is logically reversible, while the erasure process is logically irreversible in classical information theory. To see this, let us consider a classical binary measured system **S** and a binary memory **M**. As shown in Fig. 2.4a, before the measurement, the state of **M** is in the standard state "0" with unit probability, while the state of **S** is in "0" or "1." After the measurement, the state of **M** changes according to the state of **S**, and the states of **M** and **S** are perfectly correlated. In terminology of theory of computation, this process corresponds to the C-NOT gate, where **M** is the target bit. We stress that there is a one-to-one correspondence of the pre-measurement and the post-measurement states of the total system of **M** and **S**, which implies that the measurement process is logically reversible.

On the other hand, in the erasure process, measured system **S** is detached from memory **M**, and the state of **M** returns to the standard state "0" with unit probability, irrespective of the pre-erasure state. Figure 2.4b shows this process. Clearly, there is no one-to-one correspondence between the pre-erasure and the post-erasure states. In other words, the erasure process is not bijective. Therefore, the information erasure is logically irreversible.

In the logically reversible process, we may conclude that the entropy of the total state of **S** and **M** does not change because the process is reversible. This is the main reason why Bennett considered we do not need any energy cost for the measurement process in principle. On the other hand, in the logically irreversible process, the entropy may always decrease, which means that there must be an entropy increase in the environment to be consistent with the second law of thermodynamics. In Landauer's argument, this entropy increase in the environment corresponds to the heat dissipation and the work requirement for the erasure process. Therefore, according to Bennett's argument, we always need the work for the erasure process, not for the measurement process, because of the second law of thermodynamics. This argument seems to be general and fundamental, which has been accepted as the resolution of the paradox of Maxwell's demon. However, we will discuss that the logical irreversibility is in fact irrespective to the heat dissipation, and the work is not necessarily needed for information erasure.

References

1. J.C. Maxwell, *Theory of Heat* (Appleton, London, 1871)
2. L. Szilard, Z. Phys. **53**, 840 (1929)
3. H.S. Leff, A.F. Rex (eds.), *Maxwell's Demon 2: Entropy, Classical and Quantum Information, Computing* (Princeton University Press, New Jersey, 2003)
4. K. Maruyama, F. Nori, V. Vedral, Rev. Mod. Phys. **81**, 1 (2009)
5. O.J.E. Maroney, Information processing and thermodynamic entropy, in The Stanford Encyclopedia of Philosophy, ed. by E.N. Zalta (Fall 2009 Edition).
6. T. Sagawa, M. Ueda, Information thermodynamics: Maxwell's demon in nonequilibrium dynamics, in Nonequilibrium Statistical Physics of Small Systems: Fluctuation Relations and Beyond, ed. by R. Klages, W. Just, C. Jarzynski (Wiley-VCH, Weinheim, 2012), arXiv:1111.5769 (2011) (To appear).
7. L. Brillouin, J. Appl. Phys. **22**, 334 (1951)
8. R. Landauer, IBM J. Res. Dev. **5**, 183 (1961)
9. C.H. Bennett, Int. J. Theor. Phys. **21**, 905 (1982)
10. W.H. Zurek, Nature **341**, 119 (1989)
11. W.H. Zurek, Phys. Rev. A **40**, 4731 (1989)
12. K. Shizume, Phys. Rev. E **52**, 3495 (1995)
13. R. Landauer, Science **272**, 1914 (1996)
14. H. Matsueda, E. Goto, K-F. Loe, RIMS Kôkyûroku **1013**, 187 (1997)
15. B. Piechocinska, Phys. Rev. A **61**, 062314 (2000)
16. C.H. Bennett, Stud. Hist. Philos. Mod. Phys. **34**, 501 (2003)
17. A.E. Allahverdyan, T.M. Nieuwenhuizen, Phys. Rev. E **64**, 0561171 (2001)
18. C. Horhammer, H. Buttner, J. Stat. Phys. **133**, 1161 (2008)
19. M.M. Barkeshli, arXiv:cond-mat/0504323 (2005).
20. J.D. Norton, Stud. Hist. Philos. Mod. Phys. **36**, 375 (2005)
21. O.J.E. Maroney, Phys. Rev. E **79**, 031105 (2009)
22. S. Turgut, Phys. Rev. E **79**, 041102 (2009)
23. T. Sagawa, M. Ueda, Phys. Rev. Lett. **102**, 250602 (2009)
24. T. Sagawa, M. Ueda, Phys. Rev. Lett. 106, 189901(E) (2011).
25. T. Sagawa, Prog. Theor. Phys. **127**, 1 (2012)

Chapter 3
Classical Dynamics, Measurement, and Information

Abstract In this chapter, we review the basic concepts in the classical information theory [1, 2], which is needed to quantitatively discuss the relationship between thermodynamics and information. First, we formulate stochastic dynamics in classical systems. Second, we introduce the basic quantities in the classical information theory: the Shannon information, the Kullback–Leibler divergence (the relative entropy), and the mutual information. Third, we discuss classical measurement theory with stochastic errors by using the information theory. We illustrate three typical examples of classical measurements.

3.1 Classical Dynamics

We review the formulation of classical stochastic dynamics. Let **S** be a classical system and X_S be the phase space of **S**.

We first assume that X_S is a finite set. Let $P_0[x_0]$ be the probability of realizing an initial state $x_0 \in X_S$ at time 0, and $\boldsymbol{P}_0 \equiv \{P_0[x_0]\}$ be a vector whose elements are $P_0[x_0]$'s. The time evolution of the system is characterized by transition probability $P_t[x_t|x_0]$, which represents the probability of realizing $x_t \in X_S$ at time t under the condition that the system is in x_0 at time 0. We note that $\sum_{x_t} P_t[x_t|x_0] = 1$ holds. Then the probability distribution of x_t is given by

$$P_t[x_t] = \sum_{x_0} P_t[x_t|x_0]P_0[x_0]. \tag{3.1}$$

We also write Eq. (3.1) as

$$\boldsymbol{P}_t = \mathscr{E}_t(\boldsymbol{P}_0), \tag{3.2}$$

where $\mathscr{E}_t$ is a linear map on vector $\boldsymbol{P}_0$. We note that the stochastic dynamics is characterized by $\mathscr{E}_t$, or equivalently $\{P_t[x_t|x_0]\}$. The dynamics is deterministic if,

T. Sagawa, *Thermodynamics of Information Processing in Small Systems*, Springer Theses, DOI: 10.1007/978-4-431-54168-4_3,

for every x_0, there is a unique x_t that satisfies $P_t[x_t|x_0] \neq 0$. We say that the dynamics is reversible if, for every x_t, there is a unique x_0 that satisfies $P_t[x_t|x_0] \neq 0$.

We next consider the case in which X_S consists of continuous variables. The initial probability of finding the system's state in an infinitesimal interval around x_0 with width dx_0 can be written as $P_0[x_0]dx_0$, where $P_0[x_0]$ is the probability density. We also write the probability density of x_t as $P_t[x_t]$. Let $P_t[x_t|x_0]$ be the probability density of realizing $x_t \in X_S$ at time t under the condition that the system is in x_0 at time 0. Then we have

$$P_t[x_t] = \int dx_0 P_t[x_t|x_0] P_0[x_0] \tag{3.3}$$

for the case of continuous variables.

3.2 Classical Information Theory

We now shortly review the basic concepts in classical information theory [1, 2].

3.2.1 Shannon Entropy

We first consider the Shannon entropy. Let $x \in X_S$ be an arbitrary probability variable of system **S**. If x is a discrete variable whose probability distribution is $\boldsymbol{P} \equiv \{P[x]\}$, the Shannon entropy is defined as

$$H(\boldsymbol{P}) \equiv -\sum_x P[x] \ln P[x]. \tag{3.4}$$

On the other hand, if x is a continuous variable, the probability distribution is given by $\{P[x]dx\}$ and $\boldsymbol{P} \equiv \{P[x]\}$ is the set of the probability densities. In this case,

$$-\int P[x]dx \ln(P[x]dx) = -\int dx P[x] \ln P[x] - \int dx P[x] \ln(dx) \tag{3.5}$$

holds, where the second term of the right-hand side does not converge in the limit of $dx \to 0$. Therefore we define the Shannon entropy for continuous variables as

$$H(\boldsymbol{P}) \equiv -\int dx P[x] \ln P[x]. \tag{3.6}$$

We note that, for the cases of continuous variables, the Shannon entropy (3.6) is not invariant under the transformation of variable x.

We consider the condition of stochastic dynamics with which the Shannon entropy is invariant in time. For the cases of discrete variables, $H(\boldsymbol{P}_t)$ is independent of time t if the dynamics is deterministic and reversible. On the other hand, for the cases of continuous variables, the determinism and reversibility are not sufficient conditions for the time-invariance of $H(\boldsymbol{P}_t)$. In addition, we need the condition that the integral element dx_t is time-invariant, or equivalently, the phase-space volume is time-invariant. This condition is satisfied if the system obeys a Hamiltonian dynamics that satisfies Liouville's theorem.

The Shannon entropy satisfies the following important properties, which are valid for both discrete and continuous variables. For simplicity, here we discuss only discrete variables.

We first consider the case that the probability distribution $\boldsymbol{P} \equiv \{P[x]\}$ is given by the statistical mixture of other distributions $\boldsymbol{P}_k \equiv \{P_k[x]\}$ $(k = 1, 2, \ldots)$ as

$$P[x] = \sum_k q_k P_k[x], \tag{3.7}$$

where $\{q_k\} \equiv \boldsymbol{q}$ is the distribution of k's, satisfying $\sum_k q_k = 1$. Then the total Shannon entropy of $\boldsymbol{P}$ satisfies

$$\sum_k q_k H(\boldsymbol{P}_k) \leq H(\boldsymbol{P}) \leq \sum_k q_k H(\boldsymbol{P}_k) + H(\boldsymbol{q}). \tag{3.8}$$

The left equality of (3.8) is achieved if and only if all of $\boldsymbol{P}_k$'s are identical. On the other hand, the right equality of (3.8) is achieved if and only if the supports of $\boldsymbol{P}_k$'s are mutually non-crossing.

We next consider two systems $\mathbf{S}_1$ and $\mathbf{S}_2$, whose phase spaces are X_{S_1} and X_{S_2}, respectively. Let $\boldsymbol{P} \equiv \{P[x_1, x_2]\}$ be the joint probability distribution of $(x_1, x_2) \in X_{S_1} \times X_{S_2}$. The marginal distributions are given by $\boldsymbol{P}_1 \equiv \{P_1[x_1]\}$ and $\boldsymbol{P}_2 \equiv \{P_2[x_2]\}$ with $P_1[x_1] \equiv \sum_{x_2} P[x_1, x_2]$ and $P_2[x_2] \equiv \sum_{x_1} P[x_1, x_2]$. Then the Shannon entropy satisfies the subadditivity

$$H(\boldsymbol{P}) \leq H(\boldsymbol{P}_1) + H(\boldsymbol{P}_2). \tag{3.9}$$

The equality in (3.9) holds if and only if the two systems are not correlated, i.e., $P[x_1, x_2] = P_1[x_1] P_2[x_2]$.

3.2.2 Kullback–Leibler Divergence

We next consider the Kullback–Leibler divergence (or the relative entropy) [3]. Let $\boldsymbol{p} \equiv \{p[x]\}$ and $\boldsymbol{q} \equiv \{q[x]\}$ be two probability distributions of $x \in X_S$ for the case of discrete variables. Then their Kullback–Leibler divergence is given by

$$H(\boldsymbol{p} \| \boldsymbol{q}) \equiv \sum_x p[x] \ln \frac{p[x]}{q[x]}. \tag{3.10}$$

If x is a continuous variable with probability densities $\boldsymbol{p}$ and $\boldsymbol{q}$, the Kullback–Leibler divergence is given by

$$H(\boldsymbol{p}\|\boldsymbol{q}) \equiv \int dx p[x] \ln \frac{p[x]}{q[x]}, \tag{3.11}$$

which is invariant under the transformation of variable x, in contrast to the Shannon entropy.

From inequality $\ln(p/q) \geq 1 - (p/q)$, we obtain

$$\int dx p[x] \ln q[x] \leq \int dx p[x] \ln p[x], \tag{3.12}$$

which is called Klein's inequality. The equality in inequality (3.12) is achieved if and only if $p[x] = q[x]$ for every x (for discrete variables) or for almost every x (for continuous variables). Inequality (3.12) leads to

$$H(\boldsymbol{p}\|\boldsymbol{q}) \geq 0. \tag{3.13}$$

One of the most important properties of the Kullback–Leibler divergence with discrete variables is the monotonicity under stochastic dynamics, that is,

$$H(\mathscr{E}(\boldsymbol{p})\|\mathscr{E}(\boldsymbol{q})) \leq H(\boldsymbol{p}\|\boldsymbol{q}) \tag{3.14}$$

holds for an arbitrary stochastic dynamics $\mathscr{E}$. The equality in (3.14) is achieved if $\mathscr{E}$ is deterministic and reversible.

3.2.3 Mutual Information

We next consider the mutual information between two systems $\mathbf{S}_1$ and $\mathbf{S}_2$. Let X_{S_1} and X_{S_2} be phase spaces of $\mathbf{S}_1$ and $\mathbf{S}_2$, respectively. Let $\boldsymbol{P} \equiv \{P[x_1, x_2]\}$ be the joint probability distribution of $(x_1, x_2) \in X_{\mathrm{S}_1} \times X_{\mathrm{S}_2}$. The marginal distributions are given by $\boldsymbol{P}_1 \equiv \{P_1[x_1]\}$ and $\boldsymbol{P}_2 \equiv \{P_2[x_2]\}$ with $P_1[x_1] \equiv \sum_{x_2} P[x_1, x_2]$ and $P_2[x_2] \equiv \sum_{x_1} P[x_1, x_2]$. Then the mutual information is given by

$$I(\mathbf{S}_1 : \mathbf{S}_2) \equiv H(\boldsymbol{P}_1) + H(\boldsymbol{P}_2) - H(\boldsymbol{P}), \tag{3.15}$$

which represents the correlation between the two systems. Mutual information (3.15) can be rewritten as

$$I(\mathbf{S}_1 : \mathbf{S}_2) = \sum_{x_1, x_2} P[x_1, x_2] \ln \frac{P[x_1, x_2]}{P_1[x_1] P_2[x_2]} = H(\boldsymbol{P}\|\boldsymbol{P}'), \tag{3.16}$$

where $\boldsymbol{P}' \equiv \{P_1[x_1]P_2[x_2]\}$. From Eq. (3.16) and inequality (3.9), we find that the mutual information satisfies

$$I(\mathrm{S}_1 : \mathrm{S}_2) \geq 0, \tag{3.17}$$

where $I(\mathrm{S}_1 : \mathrm{S}_2) = 0$ is achieved if and only if $P[x_1, x_2] = P_1[x_1]P_2[x_2]$ holds, or equivalently, if the two systems are not correlated. We can also show that

$$0 \leq I(\mathrm{S}_1 : \mathrm{S}_2) \leq H(\boldsymbol{P}_1), \quad 0 \leq I(\mathrm{S}_1 : \mathrm{S}_2) \leq H(\boldsymbol{P}_2). \tag{3.18}$$

Here, $I(\mathrm{S}_1 : \mathrm{S}_2) = H(\boldsymbol{P}_1)$ holds if x_1 is determined only by x_2, and $I(\mathrm{S}_1 : \mathrm{S}_2) = H(\boldsymbol{P}_2)$ holds if x_2 is determined only by x_1.

We note that, for the case of continuous variables, Eq. (3.16) can be written as

$$I(\mathrm{S}_1 : \mathrm{S}_2) = \int dx_1 dx_2 P[x_1, x_2] \ln \frac{P[x_1, x_2]}{P_1[x_1]P_2[x_2]} = H(\boldsymbol{P} \| \boldsymbol{P}'), \tag{3.19}$$

which is invariant under the transformation of variables x_1 and x_2.

3.3 Classical Measurement Theory

We next review the general theory of a measurement on a classical system. Although the following argument can be applied to both continuous and discrete variables, for simplicity, we mainly concern the continuous variable cases.

Let $x \in X_{\mathrm{S}}$ be an arbitrary probability variable of measured system S, and $\boldsymbol{P} \equiv \{P[x]\}$ be the probability distribution or the probability densities of x. We perform a measurement on S and obtain outcome y. We note that y is also a probability variable. If the measurement is error-free, $x = y$ holds, in other words, x and y are perfectly correlated. In general, stochastic errors are involved in the measurement, so that the correlation between x and y is not perfect. The errors can be characterized by conditional probability $P[y|x]$, which represents the probability of obtaining outcome y under the condition that the true value of the measured system is x. We note that $\int dy P[y|x] = 1$ for all x, where the integral is replaced by the summation if x is a discrete variable. In the case of an error-free measurement, $P[y|x]$ is given by the delta function (or Kronecker's delta) such that $x = y$ holds. The joint probability of x and y is given by $P[x, y] = P[y|x]P[x]$, and the probability of obtaining y by $P[y] = \int dx P[x, y]$. The probability $P[x|y]$ of realizing x under the condition that the measurement outcome is y is given by

$$P[x|y] = \frac{P[y|x]P[x]}{P[y]}, \tag{3.20}$$

which is the Bayes theorem.

We next discuss the information contents of the measurement. The randomness of the measurement outcome is characterized by the Shannon entropy of y, to which we refer as the Shannon information. In general, if a probability variable is an outcome of a measurement, we call the corresponding Shannon entropy as the Shannon information. On the other hand, the effective information obtained by the measurement is characterized by the mutual information between x and y, which represents the correlation between the system's state and the measurement outcome.

We illustrate the following typical examples.

Example 1: Gaussian error. If the Gaussian noise is involved in the measurement, the error is characterized by

$$P[y|x] = \frac{1}{\sqrt{2\pi N}} \exp\left(-\frac{(y-x)^2}{2N}\right), \tag{3.21}$$

where N is the variance of the noise. For simplicity, we assume that the distribution of x is also Gaussian as $P[x] = (2\pi S)^{-1/2} \exp(-x^2/2S)$. The distribution y is then given by $P[y] = (2\pi(S+N))^{-1/2} \exp(-y^2/2(S+N))$. In this case, the Shannon information is given by

$$H \equiv -\int dy P[y] \ln P[y] = \frac{\ln(S+N) + \ln(2\pi) + 1}{2}, \tag{3.22}$$

which is determined by the variance of y. On the other hand, the mutual information is given by

$$I \equiv \int dx dy P[x,y] \ln \frac{P[x,y]}{P[x]P[y]} = \frac{1}{2} \ln\left(1 + \frac{S}{N}\right), \tag{3.23}$$

which is only determined by the S/N ratio.

Example 2: Piecewise error-free measurement. Let X_S be the phase space of x. We divide X_S into noncrossing regimes X_y $(y = 1, 2, \ldots)$ which satisfy $X_S = \cup_y X_y$ and $X_y \cap X_{y'} = \phi$ $(y \neq y')$ with ϕ being the empty set (Fig. 3.1a). We perform the measurement and precisely find which regime x is in. The measurement outcome is given by y. The conditional probability is given by $P[y|x] = 0$ $(x \notin S_y)$ or $P[y|x] = 1$ $(x \in S_y)$, which leads to $P[y] = \sum_{x \in S_y} P[x]$. Therefore we obtain

$$I = H = -\sum_y P[y] \ln P[y]. \tag{3.24}$$

We note that $H \leq H_x$ holds, where $H_x \equiv -\sum_x P[x] \ln P[x]$ is the Shannon information of the measured system.

Example 3: Binary symmetric channel. We assume that both x and y take 0 or 1. The conditional probabilities are given by

$$P[0|0] = P[1|1] = 1 - \varepsilon, \quad P[0|1] = P[1|0] = \varepsilon, \tag{3.25}$$

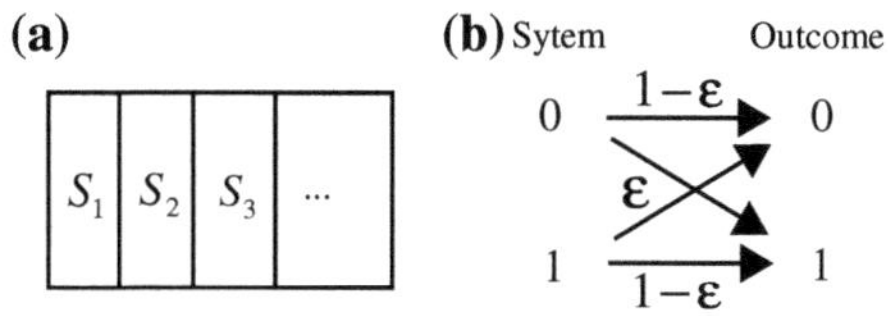

Fig. 3.1 Schematics of measurement models (reproduced from Ref. [4] with permission). **a** Piecewise error-free measurement. The total phase space is divided into subspaces $S_1, S_2, \ldots$. We measure which subspace the system is in. **b** Binary-symmetric channel with error rate ε

where ε is the error rate satisfying $0 \leq \varepsilon \leq 1$ (Fig. 3.1b). For an arbitrary probability distribution of x, the Shannon information and the mutual information are related as

$$I = H - H(\varepsilon), \tag{3.26}$$

where $H(\varepsilon) \equiv -\varepsilon \ln \varepsilon - (1-\varepsilon)\ln(1-\varepsilon)$. We note that $I = H$ holds if and only if $\varepsilon = 0$ or 1, and that $I = 0$ holds if and only if $\varepsilon = 1/2$.

References

1. C. Shannon, Bell Syst. Tech. J. 27, 379–423 and 623–656 (1948).
2. T.M. Cover, J.A. Thomas, *Elements of Information Theory* (Wiley, New York, 1991)
3. S. Kullback, R.A. Leibler, Ann. Math. Stat. **22**, 79 (1951)
4. T. Sagawa, Prog. Theor. Phys. **127**, 1 (2012)

Chapter 4
Quantum Dynamics, Measurement, and Information

Abstract In this chapter, we review the theory of quantum dynamics, measurement, and information [1]. First, we discuss quantum dynamics without measurement. In particular, we formulate the dynamics in quantum open systems by introducing Kraus operators. Second, we discuss the quantum measurement theory. In particular, we formulate quantum measurements with measurement errors by using Kraus operators (or measurement operators) and positive operator-valued measures (POVMs). Third, we discuss the basic concepts in quantum measurement theory. We introduce the von Neumann entropy, the quantum Kullback-Leibler divergence (the quantum relative entropy), and the quantum mutual information. We also introduce the QC-mutual information that plays a crucial role to formulate the generalized second law of thermodynamics with quantum feedback control. In addition, we show that the classical measurement theory is a special case of the quantum measurement theory. For simplicity, we will focus on quantum systems that correspond to finite-dimensional Hilbert spaces.

4.1 Quantum Dynamics

First of all, we discuss the theory of quantum dynamics without any measurement. We first discuss the unitary dynamics, and next the nonunitary dynamics in open systems. In the followings, we set $\hbar = 1$.

4.1.1 Unitary Evolutions

We consider a quantum system **S** corresponding to a finite-dimensional Hilbert space **H**. Let $|\psi\rangle \in \mathbf{H}$ be a pure state with $\langle\psi|\psi\rangle = 1$. If system **S** is isolated from any other quantum systems, the time evolution of state vector $|\psi\rangle$ is described

T. Sagawa, *Thermodynamics of Information Processing in Small Systems*,
Springer Theses, DOI: 10.1007/978-4-431-54168-4_4, © Springer Japan 2013

by the Schrödinger equation

$$i\frac{d}{dt}|\psi(t)\rangle = \hat{H}(t)|\psi(t)\rangle, \tag{4.1}$$

where $\hat{H}(t)$ is the Hamiltonian of this system. The formal solution of Eq. (4.1) is given by

$$|\psi(t)\rangle = \hat{U}(t)|\psi(0)\rangle, \tag{4.2}$$

where $\hat{U}(t)$ is the unitary operator that is given by

$$\hat{U}(t) \equiv T\exp\left(-\mathrm{i}\int \hat{H}(t)dt\right), \tag{4.3}$$

where T means the time-ordered product.

A statistical mixture of pure states is called a mixed state. It is described by a Hermitian operator $\hat{\rho}$ acting on **H**, which we call a density operator. The statistical mixture of pure states $\{|\xi_j\rangle\}$ with probability distribution $\{q_j\}$ satisfying $\sum_j q_j = 1$ with $q_j \geq 0$ corresponds to density operator

$$\hat{\rho} = \sum_i q_i|\xi_i\rangle\langle\xi_i|. \tag{4.4}$$

In the case of pure state $|\psi\rangle$, the corresponding density operator is given by $\hat{\rho} = |\psi\rangle\langle\psi|$. From Eq. (4.4), it can easily be shown that

$$\hat{\rho} \geq 0 \tag{4.5}$$

and

$$\mathrm{tr}(\hat{\rho}) = 1. \tag{4.6}$$

Conversely, any Hermitian operator satisfying (4.5) and (4.6) can be decomposed as

$$\hat{\rho} = \sum_i q_i|\phi_i\rangle\langle\phi_i|, \tag{4.7}$$

where $\{q_i\}$ is a probability distribution satisfying $\sum_i q_i = 1$, and $\{|\phi_i\rangle\}$ is an orthonormal basis of **H** satisfying $\langle\phi_i|\phi_j\rangle = \delta_{ij}$. The decomposition (4.7) implies that any Hermitian operator $\hat{\rho}$ that satisfies (4.5) and (4.6) can be interpreted as a statistical mixture of pure states.

From spectral decomposition (4.7), we can easily obtain the time evolution of the density operator:

$$i\frac{d}{dt}\hat{\rho}(t) = [\hat{H}(t), \hat{\rho}(t)], \tag{4.8}$$

where $[\hat{A}, \hat{B}] \equiv \hat{A}\hat{B} - \hat{B}\hat{A}$. Equation (4.8) is called the von Neumann equation. The formal solution of Eq. (4.8) is given by

$$\hat{\rho}(t) = \hat{U}(t)\hat{\rho}(0)\hat{U}(t)^\dagger, \tag{4.9}$$

where $\hat{U}(t)$ is given by Eq. (4.3). We note that the unitary evolution is trace-preserving:

$$\mathrm{tr}[\hat{U}(t)\hat{\rho}(0)\hat{U}(t)^\dagger] = \mathrm{tr}[\hat{\rho}(0)]. \tag{4.10}$$

4.1.2 Nonunitary Evolutions

We will next discuss the general formulation of quantum open systems that are subject to nonunitary evolutions.

Before that, we generally formulate the composite systems of two quantum systems $\mathbf{S}_1$ and $\mathbf{S}_2$ corresponding to the Hilbert spaces $\mathbf{H}_1$ and $\mathbf{H}_2$, respectively. Then the composite system $\mathbf{S}_1 + \mathbf{S}_2$ belongs to the Hilbert space $\mathbf{H}_1 \otimes \mathbf{H}_2$, where $\otimes$ denotes the tensor product. For the case in which $\mathbf{S}_1$ and $\mathbf{S}_2$ are not correlated, a pure state of $\mathbf{H}_1 \otimes \mathbf{H}_2$ corresponds to

$$|\Psi\rangle = |\psi_1\rangle|\psi_2\rangle \in \mathbf{H}_1 \otimes \mathbf{H}_2, \tag{4.11}$$

which is called as a separable state. If a pure state $|\Psi\rangle \in \mathbf{H}_1 \otimes \mathbf{H}_2$ cannot be factorized unlike Eq. (4.11), the state is called an entangled state. In general, a state of $\mathbf{S}_1 + \mathbf{S}_2$ is described by a density operator acting on $\mathbf{H}_1 \otimes \mathbf{H}_2$. Let $\hat{\rho}$ be a density operator of the composite system. Its marginal states $\hat{\rho}_1$ and $\hat{\rho}_2$ are defined as

$$\hat{\rho}_1 = \mathrm{tr}_2\hat{\rho} \equiv \sum_k \langle\phi_k^{(2)}|\hat{\rho}|\phi_k^{(2)}\rangle, \tag{4.12}$$

$$\hat{\rho}_2 = \mathrm{tr}_1\hat{\rho} \equiv \sum_k \langle\phi_k^{(1)}|\hat{\rho}|\phi_k^{(1)}\rangle, \tag{4.13}$$

where $\{|\phi_k^{(1)}\rangle\}$ is an arbitrary orthonormal basis of $\mathbf{H}_1$, and $\{|\phi_k^{(2)}\rangle\}$ is that of $\mathbf{H}_2$.

We now consider nonunitary evolutions of system $\mathbf{S}$ that interacts with environment $\mathbf{E}$. We note that $\mathbf{S}$ and $\mathbf{E}$ correspond to Hilbert spaces $\mathbf{H}_\mathrm{S}$ and $\mathbf{H}_\mathrm{E}$, respectively. The total system is isolated from any other quantum systems and subject to a unitary evolution. We assume that the initial state of the total system is given by a product state

$$\hat{\rho}_\mathrm{tot} \equiv \hat{\rho} \otimes |\psi\rangle\langle\psi|, \tag{4.14}$$

where the state of $\mathbf{E}$ is assumed to be described by a state vector $|\psi\rangle$. We note that the generality is not lost by this assumption, because every mixed state can be described by a vector with a sufficiently large Hilbert space. After unitary evolution $\hat{U}$ of $\mathbf{S}+\mathbf{E}$, the total state is given by

$$\hat{\rho}'_{\text{tot}} = \hat{U}\hat{\rho}_{\text{tot}}\hat{U}^\dagger, \tag{4.15}$$

which leads to **S**'s state

$$\hat{\rho}' = \text{tr}_{\text{E}}[\hat{U}\hat{\rho}_{\text{tot}}\hat{U}^\dagger]. \tag{4.16}$$

Let $\{|k\rangle\}$ be an $\mathbf{H}_{\text{E}}$'s basis. Then we have

$$\hat{\rho}' = \sum_k \langle k|\hat{U}|\psi\rangle\hat{\rho}\langle\psi|\hat{U}^\dagger|k\rangle. \tag{4.17}$$

By introducing notation

$$\hat{M}_k \equiv \langle k|\hat{U}|\psi\rangle, \tag{4.18}$$

we finally obtain

$$\hat{\rho}' = \sum_k \hat{M}_k\hat{\rho}\hat{M}_k^\dagger. \tag{4.19}$$

Equation (4.19) is called Kraus representation [2–5] and $\hat{M}_k$'s are called Kraus operators. The Kraus representation is the most basic formula to describe the dynamics of quantum open systems, which is very useful in quantum optics and quantum information theory. We note that unitary evolution can be written in the "Kraus representation" as $\hat{\rho}' = \hat{U}\hat{\rho}\hat{U}^\dagger$, where $\hat{U}$ is the single Kraus operator. We stress that Eq. (4.19) can describe nonunitary evolutions. The linear map from $\hat{\rho}$ to $\hat{\rho}'$ in Eq. (4.19) is called a quantum operation, which can be written as

$$\mathcal{E} : \hat{\rho} \mapsto \mathcal{E}(\hat{\rho}) \equiv \sum_k \hat{M}_k\hat{\rho}\hat{M}_k^\dagger. \tag{4.20}$$

We note that the Kraus operators satisfy

$$\sum_k \hat{M}_k^\dagger\hat{M}_k = \sum_k \langle\psi|\hat{U}^\dagger|k\rangle\langle k|\hat{U}|\psi\rangle = \langle\psi|\hat{I}_{\text{tot}}|\psi\rangle = \hat{I}_{\text{S}}, \tag{4.21}$$

where $\hat{I}_{\text{tot}}$ and $\hat{I}_{\text{S}}$ are the identity of $\mathbf{H}_{\text{S}} \otimes \mathbf{H}_{\text{E}}$ and $\mathbf{H}_{\text{S}}$, respectively. Equation (4.21) confirms that the trace of $\hat{\rho}$ is conserved:

$$\text{tr}[\hat{\rho}'] = \text{tr}\left[\sum_k \hat{M}_k^\dagger\hat{M}_k\hat{\rho}\right] = \text{tr}[\hat{\rho}]. \tag{4.22}$$

4.2 Quantum Measurement Theory

We next review quantum measurement theory.

4.2.1 Projection Measurement

We start with formulating the projection measurements. An observable of **S**, which is described by Hermitian operator $\hat{A}$ acting on Hilbert space **H**, can be decomposed as

$$\hat{A} = \sum_i a(i) \hat{P}_A(i), \tag{4.23}$$

where $a(i)$'s are the eigenvalues of $\hat{A}$, and $\hat{P}_A(i)$'s are projection operators satisfying $\sum_i \hat{P}_A(i) = \hat{I}$ with $\hat{I}$ being the identity operator of **H**. If we perform the projection measurement of observable $\hat{A}$ on pure state $|\psi\rangle$, then the probability of obtaining measurement outcome $a(k)$ is given by

$$p_k = \langle \psi | \hat{P}_A(k) | \psi \rangle, \tag{4.24}$$

which is called the Born rule. The corresponding post-measurement state is given by

$$|\psi_k\rangle = \frac{1}{\sqrt{p_k}} \hat{P}_A(k) |\psi\rangle, \tag{4.25}$$

which is called the projection postulate. The measurement satisfying Eqs. (4.24) and (4.25) is called the projection measurement of $\hat{A}$ [6]. We note that the average of measurement outcomes is given by

$$\langle \hat{A} \rangle \equiv \sum_k p_k a(k) = \langle \psi | \hat{A} | \psi \rangle. \tag{4.26}$$

If we perform the projection measurement of observable $\hat{A}$ on mixed state (4.7), the probability of obtaining outcome $a(k)$ is given by

$$p_k = \sum_i q_i \langle \phi_i | \hat{P}_A(k) | \phi_i \rangle = \mathrm{tr}(\hat{P}_A(k) \hat{\rho}), \tag{4.27}$$

and the post-measurement state by

$$\hat{\rho}_k = \frac{1}{p_k} \hat{P}_A(k) \hat{\rho} \hat{P}_A(k). \tag{4.28}$$

The average of measurement outcomes of observable $\hat{A}$ is given by

$$\langle \hat{A} \rangle = \mathrm{tr}(\hat{A}\hat{\rho}). \tag{4.29}$$

4.2.2 POVM and Measurement Operators

We next discuss the general formulation of quantum measurements involving measurement errors. The measurement process can be formulated by indirect measurement models, in which the measured system **S** interacts with a probe **P**. Let $\hat{\rho}$ be the measured state of **S**, and $\hat{\sigma}$ be the initial state of **P**. The initial state of the composite system is then $\hat{\rho} \otimes \hat{\sigma}$. Let $\hat{U}$ be the unitary operator which characterizes the interaction between **S** and **P** as

$$\hat{\rho} \otimes \hat{\sigma} \mapsto \hat{U}\hat{\rho} \otimes \hat{\sigma}\hat{U}^\dagger. \tag{4.30}$$

After this unitary evolution, we can extract the information about measured state $|\psi\rangle$ by performing the projection measurement of observable $\hat{R}$ of $\mathbf{S}+\mathbf{P}$. We write the spectrum decomposition of $\hat{R}$ as

$$\hat{R} \equiv \sum_i r(i)\hat{P}_R(i), \tag{4.31}$$

where $r(i) \neq r(j)$ for $i \neq j$, and $\hat{P}_R(i)$'s are projection operators with $\sum_i \hat{P}_R(i) = \hat{I}$. We stress that, in contrast to a standard textbook by Nielsen and Chuang [1], we do not necessarily assume that $\hat{R}$ is an observable of **P**, because, in some important experimental situations such as homodyne detection or heterodyne detection, $\hat{R}$ is an observable of $\mathbf{S}+\mathbf{P}$.

From the Born rule, the probability of obtaining outcome $r(i)$ is given by

$$p_k = \mathrm{tr}[\hat{P}_R(k)\hat{U}\hat{\rho} \otimes \hat{\sigma}\hat{U}^\dagger]. \tag{4.32}$$

By introducing

$$\hat{E}_k \equiv \mathrm{tr}_{\mathrm{P}}[\hat{U}^\dagger \hat{P}_R(k)\hat{U}\hat{I} \otimes \hat{\sigma}], \tag{4.33}$$

we can express p_k as

$$p_k = \mathrm{tr}[\hat{E}_k\hat{\rho}]. \tag{4.34}$$

In the case of $\hat{\sigma} = |\psi_{\mathrm{P}}\rangle\langle\psi_{\mathrm{P}}|$, Eq. (4.42) can be reduced to

$$\hat{E}_k = \langle\psi_{\mathrm{P}}|\hat{U}^\dagger \hat{P}_R(k)\hat{U}|\psi_{\mathrm{P}}\rangle. \tag{4.35}$$

The set $\{\hat{E}_k\}$ is called as a positive operator-valued measure (POVM).

We consider a special case in which $\hat{E}_k$'s are given by the projection operators $\hat{P}_A(k)$'s, which correspond to the spectral decomposition of observable $\hat{A}$ as $\hat{A} = \sum_k a(k)\hat{P}_A(k)$. In this case, the measurement can be regarded as the error-free measurement of observable $\hat{A}$. In fact, the probability distribution of the measurement outcomes obeys the Born rule in this case.

We next consider post-measurement states. Suppose that we get outcome k. Then the corresponding post-measurement state $\hat{\rho}_k$ is given by

$$\hat{\rho}_k = \mathrm{tr}_{\mathrm{P}}[\hat{P}_R(k)\hat{U}\hat{\rho}\otimes\hat{\sigma}\hat{U}^\dagger\hat{P}_R(k)]/p_k. \tag{4.36}$$

Let $\hat{\sigma} = \sum_j q_j|\psi_j\rangle\langle\psi_j|$ be the spectral decomposition with $\{|\psi_j\rangle\}$ being an orthonormal basis. Then we have

$$\hat{\rho}_k = \sum_{j,l} q_j\langle\psi_l|\hat{P}_R(k)\hat{U}|\psi_j\rangle\hat{\rho}\langle\psi_j|\hat{U}^\dagger\hat{P}_R(k)|\psi_l\rangle/p_k, \tag{4.37}$$

and define the Kraus operators as

$$\hat{M}_{k;jl} \equiv \sqrt{q_j}\langle\psi_l|\hat{P}_R(k)\hat{U}|\psi_j\rangle, \tag{4.38}$$

which is also called measurement operators in this situation. We finally have

$$\hat{\rho}_k = \frac{1}{p_k}\sum_{jl}\hat{M}_{k;jl}\hat{\rho}\hat{M}_{k;jl}^\dagger \tag{4.39}$$

and

$$\hat{E}_k = \sum_{jl}\hat{M}_{k,jl}^\dagger\hat{M}_{k,jl}. \tag{4.40}$$

If $\hat{R}$ is an observable of **R** with $\hat{P}_R(k) \equiv \sum_l |k,l\rangle\langle k,l|$, we have

$$\hat{M}_{k;jl} = \sqrt{q_j}\langle k,l|\hat{U}|\psi_j\rangle. \tag{4.41}$$

By relabeling the indexes (j,l) by j for simplicity, we summarize the formula as follows.

$$\hat{E}_k = \mathrm{tr}_{\mathrm{P}}(\hat{U}^\dagger(\hat{I}\otimes\hat{P}_R(k))\hat{U}(\hat{I}\otimes\hat{\sigma})), \tag{4.42}$$

$$p_k = \mathrm{tr}(\hat{E}_k\hat{\rho}), \tag{4.43}$$

$$\hat{\rho}_k = \frac{1}{p_k}\mathrm{tr}_{\mathrm{P}}[\hat{U}(\hat{\rho}\otimes\hat{\sigma})\hat{U}^\dagger(\hat{I}\otimes\hat{P}_R(k))] = \frac{1}{p_k}\sum_j\hat{M}_{k,j}\hat{\rho}\hat{M}_{k,j}^\dagger, \tag{4.44}$$

$$\hat{E}_k = \sum_j \hat{M}_{k,j}^{\dagger} \hat{M}_{k,j}. \tag{4.45}$$

We note that for a more special case in which $\hat{R}$ is an observable of **R** with $\hat{P}_R(k) = |k\rangle\langle k|$ for all k and $\hat{\sigma} = |\psi_{\mathrm{P}}\rangle\langle\psi_{\mathrm{P}}|$ is a pure state, Eqs. (4.41), (4.42), (4.44), and (4.45) can be simplified, respectively, as

$$\hat{M}_k = \langle k|\hat{U}|\psi_{\mathrm{P}}\rangle, \tag{4.46}$$

$$\hat{E}_k = \langle\psi_{\mathrm{P}}|\hat{U}^{\dagger}|k\rangle\langle k|\hat{U}|\psi_{\mathrm{P}}\rangle, \tag{4.47}$$

$$\hat{\rho}_k = \frac{1}{p_k}\hat{M}_k\hat{\rho}\hat{M}_k^{\dagger}, \tag{4.48}$$

$$\hat{E}_k = \hat{M}_k^{\dagger}\hat{M}_k. \tag{4.49}$$

We also note that the ensemble average of $\hat{\rho}_k$'s can be written as a trace-preserving quantum operation:

$$\sum_k p_k\hat{\rho}_k = \sum_{kj} \hat{M}_{k,j}\hat{\rho}\hat{M}_{k,j}^{\dagger}. \tag{4.50}$$

POVMs and measurement operators can be characterized by the following properties:
Positivity

$$\sum_j \hat{M}_{k,j}^{\dagger}\hat{M}_{k,j} = \hat{E}_k \geq 0, \tag{4.51}$$

Completeness

$$\sum_{kj} \hat{M}_{k,j}^{\dagger}\hat{M}_{k,j} = \sum_k \hat{E}_k = \hat{I}. \tag{4.52}$$

Equation (4.51) ensures that $p_k \geq 0$, and Eq. (4.52) ensures that $\sum_k p_k = 1$. We can show that every set of operators $\{\hat{M}_{k,j}\}$ satisfying (4.51) and (4.52) has a corresponding model of the measurement process. To see this, letting $\hat{\sigma} = |\psi_{\mathrm{P}}\rangle\langle\psi_{\mathrm{P}}|$, we define an operator $\hat{U}$ as

$$\hat{U}|\psi\rangle|\psi_{\mathrm{P}}\rangle \equiv \sum_{k,j} \hat{M}_{k,j}|\psi\rangle|\phi_{\mathrm{P}}(k,j)\rangle, \tag{4.53}$$

where $\{|\phi_{\mathrm{P}}(k,j)\rangle\}$ is an orthonormal set of the Hilbert space vectors corresponding to **P**. For arbitrary state vectors $|\psi\rangle$, $|\varphi\rangle$ of **S**, we have

$$\begin{aligned}\langle\psi|\langle\psi_{\mathrm{P}}|\hat{U}^{\dagger}\hat{U}|\varphi\rangle|\psi_{\mathrm{P}}\rangle &= \sum_{k,j,k',j'}\langle\psi|\hat{M}_{k,j}^{\dagger}\hat{M}_{k',j'}|\varphi\rangle\langle\phi_{\mathrm{P}}(k,j)|\phi_{\mathrm{P}}(k',j')\rangle \\ &= \sum_{k,j}\langle\psi|\hat{M}_{k,j}^{\dagger}\hat{M}_{k,j}|\varphi\rangle \\ &= \langle\psi|\varphi\rangle, \end{aligned} \tag{4.54}$$

where we used the completeness condition. We thus conclude that $\hat{U}$ is a unitary operator. By taking

$$\hat{P}_R(k) \equiv \hat{I}\otimes\sum_j|\phi_{\mathrm{P}}(k,j)\rangle\langle\phi_{\mathrm{P}}(k,j)|, \tag{4.55}$$

we obtain

$$\hat{M}_{k;j} \equiv \langle\phi_{\mathrm{P}}(k,j)|\hat{U}|\psi_{\mathrm{P}}\rangle \tag{4.56}$$

for all k. Therefore $\{\hat{M}_{k,j}\}$ has a model of the measurement process characterized by $\hat{U}$, $|\psi_{\mathrm{P}}\rangle$, and $\{\hat{P}_R(k)\}$. We stress that, according to the above discussion, every set of measurement operators can be realized by a measurement model for which $\hat{R}$ is the observable of **P**.

Example (Spontaneous emission of a two level atom): As a simple example, we consider a two-level atom surrounded by the vacuum in free space. We detect a photon that is spontaneously emitted from the atom. We assume that the efficiency of the detection is perfect. Let $|+\rangle \equiv [1,0]^{\mathrm{T}}$ and $|-\rangle \equiv [0,1]^{\mathrm{T}}$ be the excited and ground states, respectively. If the probability that the excited state emits a photon is p, then the Kraus operators are given by

$$\hat{M}_0 \equiv \begin{bmatrix}\sqrt{1-p} & 0\\ 0 & 1\end{bmatrix},\quad \hat{M}_1 \equiv \begin{bmatrix}0 & 0\\ \sqrt{p} & 0\end{bmatrix}. \tag{4.57}$$

Event "1" corresponds to the emission:

$$\hat{M}_1\begin{bmatrix}1\\0\end{bmatrix} = \begin{bmatrix}0\\ \sqrt{p}\end{bmatrix}. \tag{4.58}$$

If the initial state is given by

$$\hat{\rho} = \begin{bmatrix}a & b\\ b^* & 1-a\end{bmatrix}, \tag{4.59}$$

the ensemble average of the post-emission states is given by

$$\hat{\rho}' = \sum_{k=0,1}\hat{M}_k\hat{\rho}\hat{M}_k^{\dagger} = \begin{bmatrix}a(1-p) & b\sqrt{1-p}\\ b^*\sqrt{1-p} & 1-a(1-p)\end{bmatrix}. \tag{4.60}$$

We note that the non-diagonal terms are decreased by a factor of $\sqrt{1-p}$, which means that the spontaneous emission causes a decoherence. The probability that the emission occurs is given by

$$p_1 = \mathrm{tr}[\hat{M}_1^\dagger \hat{M}_1 \hat{\rho}] = ap, \tag{4.61}$$

which means that the atom is in the excited state with probability a, and if so, it emits a photon with probability p. We note that, if $p = 1$, we have

$$\hat{\rho}' = \begin{bmatrix} 0 & 0 \\ 0 & 1 \end{bmatrix}. \tag{4.62}$$

4.3 Quantum Information Theory

We now discuss the information-theoretic aspects of quantum systems. We also discuss that the classical measurement theory can be regarded a special case of the quantum measurement theory.

4.3.1 Von Neumann Entropy

We start with introducing the concept of the von Neumann entropy of density operator $\hat{\rho}$ as

$$S(\hat{\rho}) \equiv -\mathrm{tr}(\hat{\rho} \ln \hat{\rho}). \tag{4.63}$$

If $\hat{\rho}$ is diagonalized as $\hat{\rho} = \sum_k p_k |k\rangle\langle k|$, the von Neumann entropy reduces to the Shannon entropy of $\boldsymbol{p} \equiv \{p_k\}$ as

$$S(\hat{\rho}) = -\sum_k p_k \ln p_k \equiv H(\boldsymbol{p}). \tag{4.64}$$

The von Neumann entropy is invariant under an arbitrary unitary evolution:

$$S(\hat{U} \hat{\rho} \hat{U}^\dagger) = S(\hat{\rho}), \tag{4.65}$$

where $\hat{U}$ is a unitary operator. On the other hand, it increases by projections as

$$S\left(\sum_k \hat{P}_k \hat{\rho} \hat{P}_k\right) \geq S(\hat{\rho}), \tag{4.66}$$

where $\hat{P}_k$'s are projection operators satisfying $\sum_k \hat{P}_k = \hat{I}$.

The von Neumann entropy has important properties as follows.

We first suppose that $\hat{\rho}$ is a statistical mixture of $\hat{\rho}_k$'s as

$$\hat{\rho} = \sum_k p_k \hat{\rho}_k. \tag{4.67}$$

Then the following inequalities are satisfied:

$$\sum_k p_k S(\hat{\rho}_k) \le S(\hat{\rho}) \le \sum_k p_k S(\hat{\rho}_k) + H(\boldsymbol{p}), \tag{4.68}$$

where $H(\boldsymbol{p}) \equiv -\sum p_k \ln p_k$ is the Shannon entropy of the statistical mixture. The left equality of (4.68) is achieved if and only if all of $\hat{\rho}_k$'s are identical. On the other hand, the right equality of (4.68) is achieved if and only if the supports of $\hat{\rho}_k$'s are mutually orthogonal.

We next consider the composite system of $\mathbf{S}_1$ and $\mathbf{S}_2$. Let $\hat{\rho}_{12}$, $\hat{\rho}_1 \equiv \mathrm{tr}_2[\hat{\rho}_{12}]$, and $\hat{\rho}_2 \equiv \mathrm{tr}_1[\hat{\rho}_{12}]$ be the density operator of the total system, that of $\mathbf{S}_1$, and that of $\mathbf{S}_2$, respectively. Then the subadditivity is satisfied:

$$S(\hat{\rho}_{12}) \le S(\hat{\rho}_1) + S(\hat{\rho}_2), \tag{4.69}$$

where the equality is achieved if and only if $\hat{\rho}_{12} = \hat{\rho}_1 \otimes \hat{\rho}_2$.

4.3.2 Quantum Kullback-Leibler Divergence

We next discuss the quantum version of the Kullback-Leibler divergence (or the quantum relative entropy) of two quantum states $\hat{\rho}$ and $\hat{\sigma}$, which is defined as

$$S(\hat{\rho}\|\hat{\sigma}) \equiv \mathrm{tr}(\hat{\rho} \ln \hat{\rho}) - \mathrm{tr}(\hat{\rho} \ln \hat{\sigma}). \tag{4.70}$$

We can show that the quantum version of Klein's inequality (3.12):

$$\mathrm{tr}[\hat{\rho} \ln \hat{\sigma}] \le \mathrm{tr}[\hat{\rho} \ln \hat{\rho}], \tag{4.71}$$

where the equality is achieved if and only if $\hat{\rho} = \hat{\sigma}$. Inequality (4.71) leads to

$$S(\hat{\rho}\|\hat{\sigma}) \ge 0. \tag{4.72}$$

On the other hand, we consider density operators $\hat{\rho}_{12}$ and $\hat{\sigma}_{12}$ of the composite system of $\mathbf{S}_1$ and $\mathbf{S}_2$. Let $\hat{\rho}_1 \equiv \mathrm{tr}_2[\hat{\rho}_{12}]$ and $\hat{\sigma}_1 \equiv \mathrm{tr}_2[\hat{\sigma}_{12}]$. Then the following inequality holds:

$$S(\hat{\rho}_1\|\hat{\sigma}_1) \le S(\hat{\rho}_{12}\|\hat{\sigma}_{12}). \tag{4.73}$$

Combining inequality (4.73) with the unitary invariance of von Neumann entropy, we obtain

$$S(\mathscr{E}(\hat{\rho}_1)\|\mathscr{E}(\hat{\sigma}_1)) \leq S(\hat{\rho}_1\|\hat{\sigma}_1) \tag{4.74}$$

for an arbitrary quantum operation $\mathscr{E}$ on $\mathbf{S}_1$.

We note that the quantum mutual information between two systems $\mathbf{S}_1$ and $\mathbf{S}_2$ is given by

$$\begin{aligned} I(\hat{\rho}_1 : \hat{\rho}_2) &\equiv S(\hat{\rho}_1) + S(\hat{\rho}_2) - S(\hat{\rho}_{12}) && (4.75)\\ &= S(\hat{\rho}_{12}\|\hat{\rho}_1 \otimes \hat{\rho}_2) && (4.76)\\ &\geq 0. && (4.77) \end{aligned}$$

We note that $I(\hat{\rho}_1 : \hat{\rho}_2) = 0$ holds if and only if the two systems are not correlated, so that $\hat{\rho}_{12} \equiv \hat{\rho}_1 \otimes \hat{\rho}_2$.

4.3.3 Holevo Bound

Another important quantity related to the mutual information is "the Holevo χ quantity." Let $X \equiv \{x\}$ be a set of classical probability variables, and $\{\hat{\rho}_x\}_{x\in X}$ be a set of density operators that are not necessarily mutually orthogonal. Let $\hat{\rho} = \sum_x p_x \hat{\rho}_x$ with $\boldsymbol{p} \equiv \{p_k\}$ being a probability distribution. Then the Holevo χ quantity is given by

$$\chi \equiv S(\hat{\rho}) - \sum_k p_k S(\hat{\rho}_k). \tag{4.78}$$

The Holevo bound is formulated as follows. An agent called Alice prepares a state $\hat{\rho}_k$ and send it to another agent called Bob. Bob performs a quantum measurement on the system with POVM $\{\hat{E}_y\}_{y\in Y}$, where Y is the set of measurement outcomes. The joint distribution of X and Y is given by

$$p(x, y) = \mathrm{tr}[\hat{E}_y \hat{\rho}_x] p_x, \tag{4.79}$$

which gives the marginal distribution of y as

$$q(y) \equiv \sum_x p(x, y) = \mathrm{tr}[\hat{E}_y \hat{\rho}]. \tag{4.80}$$

Let $\boldsymbol{q} \equiv \{q(y)\}$. Then the classical mutual information between X and Y is given by

$$I(X : Y) = \sum_{x,y} p(x, y) \ln \frac{p(x, y)}{p(x)q(y)}. \tag{4.81}$$

The Holevo bound states that the classical mutual information is bounded by the Holevo χ quantity as

$$I(X:Y) \leq \chi, \tag{4.82}$$

which implies that the accessible information that is coded on the quantum states is bounded by χ. We note that, from inequality (4.68), the Holevo χ quantity is bounded by the Shannon information of X as $\chi \leq H(\boldsymbol{p})$, where the equality is achieved if and only if the supports of $\hat{\rho}_k$'s are mutually orthogonal. Since the non-orthogonality of density operators characterizes their indistinguishability, we can conclude that mutual information $I(X:Y)$ decreases by the indistinguishability of the quantum states.

4.3.4 QC-Mutual Information

We next discuss "QC-mutual information," which will play crucial roles in the generalizations of the second law of thermodynamics. Here, "QC" means "quantum-classical"; as we will see later, QC-mutual information characterizes a kind of correlation between a quantum system and a classical system. The QC-mutual information was first introduced by Groenewold and Ozawa [7, 8], and has been studied in Refs. [9–12].

We consider density operator $\hat{\rho}$ of quantum system **S**, and perform a quantum measurement on it. Let $\{E_y\}_{y \in Y}$ be the POVM of the measurement, where Y is the set of measurement outcomes. The probability of obtaining y is given by $p(y) = \mathrm{tr}[\hat{E}_y \hat{\rho}]$. Let $\boldsymbol{p} \equiv \{p(y)\}$ and $H(\boldsymbol{p}) \equiv -\sum_y p(y) \ln p(y)$. The QC-mutual information associated the POVM is then defined as

$$I_{\mathrm{QC}} \equiv S(\hat{\rho}) + H(\boldsymbol{p}) + \sum_y \mathrm{tr}\left(\sqrt{\hat{E}_y}\hat{\rho}\sqrt{\hat{E}_y} \ln \sqrt{\hat{E}_y}\hat{\rho}\sqrt{\hat{E}_y}\right). \tag{4.83}$$

We note that, the QC-mutual information can be rewritten as

$$I_{\mathrm{QC}} \equiv S(\hat{\rho}) - \sum_y p(y) S(\hat{\rho}^{(y)}), \tag{4.84}$$

where

$$\hat{\rho}^{(y)} \equiv \frac{1}{p(y)} \sqrt{\hat{E}_y}\hat{\rho}\sqrt{\hat{E}_y}. \tag{4.85}$$

In Refs. [10, 12], it has been shown that the QC-mutual information satisfies

$$0 \leq I_{\mathrm{QC}} \leq H(\boldsymbol{p}). \tag{4.86}$$

Here, $I_{\rm QC} = 0$ holds for all state $\hat{\rho}$ if and only if $\hat{E}_y$ is proportional to the identity operator for every y, which means that we cannot obtain any information about the system by this measurement. On the other hand, $I_{\rm QC} = H(\boldsymbol{q})$ for some $\hat{\rho}$ holds if and only if $\hat{E}_y$ is the projection operator satisfying $[\hat{\rho}, \hat{E}_y] = 0$ for every y, which means that the measurement on state $\hat{\rho}$ is classical and error-free.

The proof of inequality (4.86) is as follows. We first note that

$$-\sum_y \mathrm{tr}\left(\sqrt{\hat{E}_y}\hat{\rho}\sqrt{\hat{E}_y}\ln\sqrt{\hat{E}_y}\hat{\rho}\sqrt{\hat{E}_y}\right) = \sum_y p(y) S\left(\frac{1}{p(y)}\sqrt{\hat{E}_y}\hat{\rho}\sqrt{\hat{E}_y}\right) + H(\boldsymbol{p}), \tag{4.87}$$

where $S(\cdot)$ denotes the von Neumann entropy. We introduce auxiliary system **R** which is spanned by orthonormal basis $\{|\phi_y\rangle\}_{y\in Y}$, and note that

$$S\left(\frac{1}{p(y)}\sqrt{\hat{E}_y}\hat{\rho}\sqrt{\hat{E}_y}\right) = S\left(\frac{1}{p(y)}\sqrt{\hat{E}_y}\hat{\rho}\sqrt{\hat{E}_y}\otimes|\phi_y\rangle\langle\phi_y|\right). \tag{4.88}$$

Noting that $S(\hat{L}^\dagger\hat{L}) = S(\hat{L}\hat{L}^\dagger)$ holds for any linear operator $\hat{L}$, we have

$$\begin{aligned} -\sum_y \mathrm{tr}\left(\sqrt{\hat{E}_y}\hat{\rho}\sqrt{\hat{E}_y}\ln\sqrt{\hat{E}_y}\hat{\rho}\sqrt{\hat{E}_y}\right) &= \sum_y p(y) S\left(\frac{1}{p(y)}\sqrt{\hat{E}_y}\hat{\rho}\sqrt{\hat{E}_y}\otimes|\phi_y\rangle\langle\phi_y|\right) + H(\boldsymbol{p}) \\ &= \sum_y p(y) S\left(\frac{1}{p(y)}\sqrt{\hat{\rho}}\hat{E}_y\sqrt{\hat{\rho}}\otimes|\phi_y\rangle\langle\phi_y|\right) + H(\boldsymbol{p}). \end{aligned} \tag{4.89}$$

Since $\sqrt{\hat{\rho}}\hat{E}_y\sqrt{\hat{\rho}}\otimes|\phi_y\rangle\langle\phi_y|/p(y)$'s are mutually orthogonal, we have

$$\sum_y p(y) S\left(\frac{1}{p(y)}\sqrt{\hat{\rho}}\hat{E}_y\sqrt{\hat{\rho}}\otimes|\phi_y\rangle\langle\phi_y|\right) + H(\boldsymbol{p}) = S(\hat{\sigma}), \tag{4.90}$$

where

$$\hat{\sigma} \equiv \sum_y \sqrt{\hat{\rho}}\hat{E}_y\sqrt{\hat{\rho}}\otimes|\phi_y\rangle\langle\phi_y|. \tag{4.91}$$

We note that $\mathrm{tr}_{\rm R}(\hat{\sigma}) = \hat{\rho}$ and $\mathrm{tr}_{\rm S}(\hat{\sigma}) = \sum_y p(y)|\phi_y\rangle\langle\phi_y| \equiv \hat{\rho}_{\rm R}$ hold. Therefore

$$\begin{aligned} -\sum_y \mathrm{tr}\left(\sqrt{\hat{E}_y}\hat{\rho}\sqrt{\hat{E}_y}\ln\sqrt{\hat{E}_y}\hat{\rho}\sqrt{\hat{E}_y}\right) &= S(\hat{\sigma}) \\ &\le S(\hat{\rho}) + S(\hat{\rho}_{\rm R}) \\ &= S(\hat{\rho}) + H(\boldsymbol{p}), \end{aligned} \tag{4.92}$$

which implies $I_{QC} \geq 0$. The equality in (4.92) holds for every $\hat{\rho}$ if and only if $\hat{\sigma}$ can be written as tensor product $\hat{\rho} \otimes \hat{\rho}_R$ for every $\hat{\rho}$: that is, $\hat{E}_y$ is proportional to the identity operator for all y.

We will next show that $I_{QC} \leq H(\boldsymbol{p})$. We first note that

$$H(\boldsymbol{p}) - I_{QC} = H(\boldsymbol{p}) + \sum_y p(y) S(\hat{\rho}^{(y)}) - S(\hat{\rho}). \tag{4.93}$$

Let $\hat{\rho}' \equiv \sum_y p(y)\hat{\rho}^{(y)} = \sum_y \sqrt{\hat{E}_y}\hat{\rho}\sqrt{\hat{E}_y}$. We make the spectral decompositions as $\hat{\rho} = \sum_i r_i |\psi_i\rangle\langle\psi_i|$ and $\hat{\rho}' = \sum_j s_j |\psi'_j\rangle\langle\psi'_j|$, where $s_j = \sum_i r_i d_{ij}$, and define $e_{ij} \equiv \sum_y |\langle\psi_i|\sqrt{\hat{E}_y}|\psi'_j\rangle|^2$, where $\sum_i e_{ij} = 1$ for all j and $\sum_j e_{ij} = 1$ for all i. It follows from the convexity of $-x \ln x$ that $S(\hat{\rho}) = -\sum_i r_i \ln r_i \leq -\sum_j s_j \ln s_j = S(\hat{\rho}')$. Therefore,

$$\begin{aligned} H(\boldsymbol{p}) - I_{QC} &= H(\boldsymbol{p}) + \sum_y p(y) S(\hat{\rho}^{(y)}) - S(\hat{\rho}) \\ &\geq H(\boldsymbol{p}) + \sum_k p(y) S(\hat{\rho}^{(y)}) - S(\hat{\rho}') \\ &\geq 0. \end{aligned} \tag{4.94}$$

The necessary and sufficient conditions that $H(\boldsymbol{p}) = I_{QC}$ for a given $\hat{\rho}$ are:

- $\hat{E}_y$ is a projection operator on the support of $\hat{\rho}$ for every y.
- $[\hat{\rho}, \hat{E}_y] = 0$ for every y.

We next discuss another inequality of the QC-mutual information. Let $\hat{M}_{yi}$'s be the measurement operators, which leads to an element of the POVM as $\hat{E}_y \equiv \sum_i \hat{M}^\dagger_{yi}\hat{M}_{yi}$. Let $p_{yi} \equiv \mathrm{tr}[\hat{\rho}\hat{M}^\dagger_{yi}\hat{M}_{yi}]$ which satisfies $p(y) = \sum_i p_{yi}$. The corresponding QC-mutual information is given by Eq. (4.83) or (4.84). On the other hand, we define a different POVM whose elements are given by $\hat{E}'_{yi} \equiv \hat{M}^\dagger_{yi}\hat{M}_{yi}$. The QC-mutual information corresponding to POVM $\{\hat{E}'_{yi}\}$ is then given by

$$I'_{QC} \equiv S(\hat{\rho}) - \sum_y p_{yi} S\left(\sqrt{\hat{E}'_{yi}}\hat{\rho}\sqrt{\hat{E}'_{yi}}/p_{yi}\right) = S(\hat{\rho}) - \sum_y p_{yi} S\left(\hat{M}^\dagger_{yi}\hat{\rho}\hat{M}_{yi}/p_{yi}\right). \tag{4.95}$$

Noting that

$$\sum_i \frac{p_{yi}}{p(y)} S(\hat{M}_{yi}\hat{\rho}\hat{M}_{yi}^\dagger / p_{yi}) = \sum_i \frac{p_{yi}}{p(y)} S\left(\sqrt{\hat{\rho}}\hat{M}_{yi}^\dagger \hat{M}_{yi}\sqrt{\hat{\rho}}/p_{yi}\right)$$
$$\leq S\left(\sqrt{\hat{\rho}}\hat{E}_y\sqrt{\hat{\rho}}/p(y)\right)$$
$$= S\left(\sqrt{\hat{E}_y}\hat{\rho}\sqrt{\hat{E}_y}/p(y)\right), \tag{4.96}$$

we obtain

$$I_{\rm QC} \leq I'_{\rm QC}. \tag{4.97}$$

Inequality (4.97) implies that the QC-mutual information decreases by the coarse-graining.

We next discuss the relationship between the QC-mutual information and the Holevo χ quantity. For simplicity, we consider the case in which the measurement operators are written as $\hat{M}_y = \sqrt{\hat{E}_y}$ where $\hat{E}_k$'s are the elements of the POVM. In this case, the post-measurement state with outcome y is given by $\hat{\rho}^{(y)} \equiv \hat{M}_y\hat{\rho}\hat{M}_y/p(y)$, where $p(y) \equiv {\rm tr}[\hat{E}_y\hat{\rho}]$. Let $\hat{\rho} \equiv \sum_y \hat{M}_y\hat{\rho}\hat{M}_y$. Then the QC-mutual information can be written as

$$I_{\rm QC} = \chi - \Delta S_{\rm meas}, \tag{4.98}$$

where

$$\chi \equiv S(\hat{\rho}') - \sum_y p_k S(\hat{\rho}^{(y)}) \tag{4.99}$$

is the Holevo χ quantity of the post-measurement states $\{\hat{\rho}^{(y)}\}$, and

$$\Delta S_{\rm meas} \equiv S(\hat{\rho}') - S(\hat{\rho}) \tag{4.100}$$

is the difference in the von Neumann entropy between the pre-measurement and post-measurement states. If $\Delta S_{\rm meas} = 0$ holds, that is, if the measurement process does not disturb the measured system, then $I_{\rm QC}$ reduces to the Holevo χ quantity.

4.3.5 Quantum-Classical Correspondence

We now show that the classical measurement theory discussed in Chap. 3 is a special case of the quantum measurement theory. We write the classical probability distribution as $\boldsymbol{p} \equiv (p[1], p[2], \ldots, p[n])$, where $1, 2, \ldots, n$ denote the states of the measured system. The classical distribution $\boldsymbol{p}$ corresponds to a diagonal density operator $\hat{\rho} \equiv {\rm diag}(p_1, p_2, \ldots, p_n)$, where ${\rm diag}(\cdots)$ means the diagonal matrix whose diagonal elements are given by "$\cdots$." On the other hand, for every measurement outcome

$y(= 1, 2, \cdots, m)$, the conditional probabilities $p[y|x]$ $(x = 1, 2, \ldots, n)$ correspond to a diagonal measurement operator $\hat{M}_y \equiv \mathrm{diag}(\sqrt{p[y|1]}, \sqrt{p[y|2]}, \ldots, \sqrt{p[y|n]})$. Then the POVM is given by $\hat{E}_y \equiv \hat{M}_y^\dagger \hat{M}_y = \mathrm{diag}(p[y|1], p[y|2], \ldots, p[y|n])$, which commutes with the measured density operator. We note that the probability of obtaining outcome y is given by

$$q[y] \equiv \sum_x p[y|x]p[x] = \mathrm{tr}[\hat{E}_y \hat{\rho}]. \tag{4.101}$$

The joint distribution of x and y corresponds to

$$\hat{M}_y \hat{\rho} \hat{M}_y^\dagger = \mathrm{diag}(p[y|1]p[1], p[y|2]p[2], \ldots, p[y|n]p[n]). \tag{4.102}$$

Therefore we obtain the post-measurement state with outcome y:

$$\frac{1}{q[y]} \hat{M}_y \hat{\rho} \hat{M}_y^\dagger = \mathrm{diag}\left(\frac{p[y|1]p[1]}{q[y]}, \frac{p[y|2]p[2]}{q[y]}, \ldots, \frac{p[y|n]p[n]}{q[y]} \right), \tag{4.103}$$

which corresponds the Bayes theorem (3.20). We note that, in the cases of classical measurements, every element of a POVM is always written as $\hat{E}_y = \hat{M}_y^\dagger \hat{M}_y$ with $\hat{M}_y$ being a measurement operator. We also note that

$$\sum_y \hat{M}_y \hat{\rho} \hat{M}_y^\dagger = \hat{\rho} \tag{4.104}$$

holds for classical measurement, which implies that we can neglect the effect of the decoherence due to the measurement in the cases of classical measurements.

We next show that the QC-mutual information reduces to the classical mutual information in the case of the classical measurements. In this case, we have

$$S(\hat{\rho}) = -\sum_x p[x] \ln p[x] \tag{4.105}$$

and

$$-\sum_y \mathrm{tr}\left(\sqrt{\hat{E}_y} \hat{\rho} \sqrt{\hat{E}_y} \ln \sqrt{\hat{E}_y} \hat{\rho} \sqrt{\hat{E}_y} \right) = -\sum_{x,y} p[y|x]p[x] \ln p[y|x]p[x]. \tag{4.106}$$

Therefore I_{QC} can be written as

$$I_{\mathrm{QC}} = -\sum_y q[y] \ln q[y] - \sum_x p[x] \ln p[x] + \sum_{x,y} p[y|x]p[x] \ln p[y|x]p[x], \tag{4.107}$$

which is the classical mutual information.

Therefore, the arguments based on quantum theory in the following chapters involve the classical cases, by regarding classical measurements and the classical mutual information as special cases of quantum measurements and the QC-mutual information.

References

1. M.A. Nielsen, I.L. Chuang, *Quantum Computation and Quantum Information* (Cambridge University Press, Cambridge, 2000)
2. E.B. Davies, J.T. Lewis, Commun. Math. Phys. **17**, 239 (1970)
3. K. Kraus, Ann. Phys. **64**, 311 (1971)
4. M.-D. Choi, Linear Algebra Appl. **10**, 285 (1975)
5. M. Ozawa, J. Math. Phys. **25**, 79 (1984)
6. J. von Neumann, Mathematische Grundlagen der Quantumechanik (Springer, Berlin, 1932). [Eng. trans. R. T. Beyer, Mathematical Foundations of Quantum Mechanics (Prinston University Press, Princeton, 1955)].
7. H.J. Groenewold, Int. J. Theor. Phys. **4**, 327 (1971)
8. M. Ozawa, J. Math. Phys. **27**, 759 (1986)
9. F. Buscemi, M. Hayashi, M. Horodecki, Phys. Rev. Lett. **100**, 210504 (2008)
10. T. Sagawa, M. Ueda, Phys. Rev. Lett. **100**, 080403 (2008)
11. T. Sagawa, Prog. Theor. Phys. **127**, 1 (2012)
12. T. Sagawa, Second Law-Like Inequalities with Quantum Relative Entropy: An Introduction, to appear in Lectures on Quantum Computing, Thermodynamics and Statistical Physics, Kinki University Series on Quantum Computing (World Scientific, New Jersey, 2012); e-print: arXiv:1202.0983.

Chapter 5
Unitary Proof of the Second Law of Thermodynamics

Abstract In this chapter, we review how to derive the second law of thermodynamics for systems that obey quantum mechanics at the microscopic level. Starting with the statement of the second law, we derive it based on quantum statistical mechanics [1–5]. We formulate the theory such that the total system of the thermodynamic system and the heat baths obey the unitary evolution, and assume that the initial states of the heat baths are in the canonical distribution. Mathematically, our derivation is based on the Klein inequality (or equivalently, the positivity of the quantum relative entropy). The reason why the second law can be derived from the reversible unitary evolution is due to the fact that we select the canonical distributions as the initial states.

5.1 Second Law of Thermodynamics

Since the Nineteenth century [6], the second law of thermodynamics has been established for macroscopic systems. From the modern point of view, there are several expressions of the second law [7–12]. In particular, if a macroscopic thermodynamic system **S** is in contact with a large single heat bath at temperature $T = (k_B\beta)^{-1}$, the second law for isothermal processes is formulated as follows. Suppose that the initial state of **S** is in thermal equilibrium at temperature T. We then perform thermodynamic operations on **S** through external parameters such as the volume of the gas. We do not assume that **S** is in thermal equilibrium during the operation. After that, system **S** goes back to a thermal equilibrium at temperature T. In this case, the work $W^{\rm S}$ performed on **S** is bounded by the difference of the Helmholtz free energy $\Delta F^{\rm S}$ as

$$W^{\rm S} \geq \Delta F^{\rm S}. \tag{5.1}$$

We stress that, in general, inequality (5.1) holds even if the intermediate states of the thermodynamic operation are out of equilibrium. The equality of (5.1) is achieved if

T. Sagawa, *Thermodynamics of Information Processing in Small Systems*, Springer Theses, DOI: 10.1007/978-4-431-54168-4_5, © Springer Japan 2013

the process is quasi-static, i.e., if all of the intermediate states are in thermal equilibrium. In the case of a thermodynamic cycle, inequality (5.1) reduces to Kelvin's principle:

$$W_{\rm ext}^{\rm S} \leq 0, \tag{5.2}$$

where $W_{\rm ext}^{\rm S} \equiv -W^{\rm S}$. We note that, according to thermodynamics, the Helmholtz free energy $F^{\rm S}$ and the internal energy $E^{\rm S}$ are related to the thermodynamic entropy $S_{\rm therm}^{\rm S}$ as

$$S_{\rm therm}^{\rm S} = \beta(E^{\rm S} - F^{\rm S}). \tag{5.3}$$

We next suppose that thermodynamic system **S** can contact multi-heat baths $\mathbf{B}_1, \mathbf{B}_2, \ldots, \mathbf{B}_n$, at respective temperatures $T_1 = (k_{\rm B}\beta_1)$, $T_2 = (k_{\rm B}\beta_2), \ldots, T_n = (k_{\rm B}\beta_n)$. If the initial and final states of **S** are in thermal equilibrium at temperature T and the process is a thermodynamic cycle, then the second law of thermodynamics is expressed as the Clausius inequality:

$$\sum_m \beta_m Q_m \leq 0, \tag{5.4}$$

where Q_m is the heat absorbed by **S** from $\mathbf{B}_m$. If there are only two heat baths $\mathbf{B}_{\rm H}$ and $\mathbf{B}_{\rm L}$ at respective temperatures $T_{\rm H}$ and $T_{\rm L}$, inequality (5.4) gives the Carnot bound:

$$\frac{W_{\rm ext}^{\rm S}}{Q_{\rm H}} = \frac{Q_{\rm H} + Q_{\rm L}}{Q_{\rm H}} \leq 1 - \frac{T_{\rm L}}{T_{\rm H}}, \tag{5.5}$$

where $Q_{\rm H}$ ($Q_{\rm L}$) is the heat absorbed by **S** from $\mathbf{B}_{\rm H}$ ($\mathbf{B}_{\rm L}$), and $W_{\rm ext}^{\rm S} = Q_{\rm H} + Q_{\rm L}$ is the work that is extracted from **S**.

In terms of statistical mechanics, thermodynamic quantities in thermal equilibrium can be calculated by using probability models. One of the most useful probability models is the canonical distribution:

$$\hat{\rho}_{\rm can}^{\rm S} \equiv \frac{e^{-\beta \hat{H}^{\rm S}}}{Z^{\rm S}}, \tag{5.6}$$

where $\hat{H}^{\rm S}$ is the Hamiltonian of the system, and

$$Z^{\rm S} \equiv {\rm tr}[e^{-\beta \hat{H}^{\rm S}}]. \tag{5.7}$$

With the canonical distribution, free energy $F^{\rm S}$ can be calculated as

$$F^{\rm S} = -k_{\rm B} T \ln Z^{\rm S}, \tag{5.8}$$

and internal energy $E^{\rm S}$ as

$$E^{\rm S} = {\rm tr}[\hat{H}^{\rm S} \hat{\rho}_{\rm can}^{\rm S}]. \tag{5.9}$$

From Eqs. (5.8) and (5.9), we obtain

$$\beta(E^{\rm S} - F^{\rm S}) = -{\rm tr}[\hat{\rho}^{\rm S}_{\rm can} \ln \hat{\rho}^{\rm S}_{\rm can}] \equiv S(\hat{\rho}^{\rm S}_{\rm can}), \tag{5.10}$$

where $S(\cdots)$ is the von Neumann entropy. Combining Eqs. (5.3) and (5.10), we obtain

$$S^{\rm S}_{\rm therm} = S(\hat{\rho}^{\rm S}_{\rm can}), \tag{5.11}$$

which implies that the thermodynamic entropy and the von Neumann entropy are equivalent in the canonical distribution. We note that, however, this equivalence rigorously holds only for the canonical distribution.

In the following arguments of this thesis, we sometimes assume that the initial distribution of a thermodynamic system is in the canonical distribution, while we do not assume that any intermediate or final state is in the canonical distribution. In such cases, we assume the equivalence between the von Neumann entropy and the thermodynamic entropy only in the initial state.

In this chapter, we will discuss the following two problems:

- How to prove the second law?
- How to generalize the second law to microscopic systems?

We will answer theses questions by formulating the total system of the thermodynamic system and heat baths as a unitary system. The crucial assumption is that the total system (or at least the heat baths) is in the canonical distribution in the initial state. This formulation is a standard method to prove the second law of thermodynamics and the nonequilibrium equalities in nonequilibrium statistical mechanics. Since our formulation does not involve any factor to characterize the size of the system, our proof can be applied to microscopic systems.

5.2 Initial Canonical Distribution with a Single Heat Bath

We first consider a quantum system that obeys a unitary evolution from time 0 to τ. The Hamiltonian of the system is given by $\hat{H}(\lambda)$, where λ describes a set of external parameters such as an applied magnetic field or the volume of the gas. We control λ from time 0 to τ with time-dependent protocol $\lambda(t)$. Let $\hat{H}_{\rm i} \equiv \hat{H}(\lambda(0))$ and $\hat{H}_{\rm f} \equiv \hat{H}(\lambda(\tau))$. We define the partition functions and the Helmholtz free energies with temperature $T = (k_{\rm B}\beta)^{-1}$corresponding to the initial and final Hamiltonians:

$$Z_{\rm i} \equiv {\rm tr}[e^{-\beta \hat{H}_{\rm i}}], \quad Z_{\rm f} \equiv {\rm tr}[e^{-\beta \hat{H}_{\rm f}}], \tag{5.12}$$

and

$$F_{\rm i} \equiv -k_{\rm B}T \ln Z_{\rm i}, \quad F_{\rm f} \equiv -k_{\rm B}T \ln Z_{\rm f}. \tag{5.13}$$

The initial state of the system is assumed to be the canonical distribution at temperature T:

$$\hat{\rho}_{\mathrm{i}} = \hat{\rho}_{\mathrm{can,i}} \equiv \frac{e^{-\beta \hat{H}_{\mathrm{i}}}}{Z_{\mathrm{i}}}. \tag{5.14}$$

The system evolves with the unitary evolution

$$\hat{U} \equiv T \exp\left(-\mathrm{i} \int_0^\tau \hat{H}(\lambda(t))dt\right), \tag{5.15}$$

where T means the time-ordered product. Then the final state of the system is given by

$$\hat{\rho}_{\mathrm{f}} = \hat{U}\hat{\rho}_{\mathrm{i}}\hat{U}^\dagger, \tag{5.16}$$

which is not necessarily equal to the canonical distribution $\hat{\rho}_{\mathrm{can,f}} \equiv e^{-\beta \hat{H}_{\mathrm{f}}}/Z_{\mathrm{f}}$.

Since the von Neumann entropy $S(\cdot)$ is time-invariant under unitary evolutions, we obtain

$$S(\hat{\rho}_{\mathrm{i}}) = S(\hat{\rho}_{\mathrm{f}}). \tag{5.17}$$

On the other hand, from Klein's inequality, we have

$$S(\hat{\rho}_{\mathrm{f}}) \le -\mathrm{tr}[\hat{\rho}_{\mathrm{f}} \ln \hat{\rho}_{\mathrm{can,f}}], \tag{5.18}$$

where the equality is achieved if and only if the final state is in the canonical distribution: $\hat{\rho}_{\mathrm{f}} = \hat{\rho}_{\mathrm{can,f}}$. We can also show that

$$S(\hat{\rho}_{\mathrm{i}}) = \beta(\mathrm{tr}[\hat{H}_{\mathrm{i}}\hat{\rho}_{\mathrm{i}}] - F_{\mathrm{i}}) \tag{5.19}$$

because the system is initially in the canonical distribution (i.e., $\hat{\rho}_{\mathrm{i}} = \hat{\rho}_{\mathrm{i,can}}$), and that

$$-\mathrm{tr}[\hat{\rho}_{\mathrm{f}} \ln \hat{\rho}_{\mathrm{can,f}}] = \beta(\mathrm{tr}[\hat{H}_{\mathrm{f}}\hat{\rho}_{\mathrm{f}}] - F_{\mathrm{f}}). \tag{5.20}$$

Therefore we obtain

$$\mathrm{tr}[\hat{H}_{\mathrm{f}}\hat{\rho}_{\mathrm{f}}] - \mathrm{tr}[\hat{H}_{\mathrm{i}}\hat{\rho}_{\mathrm{i}}] \ge F_{\mathrm{f}} - F_{\mathrm{i}}. \tag{5.21}$$

We note that the left-hand side of (5.21) is the difference of the energies of the initial and final states. Since the system is not in contact with another heat bath, we can identify the energy difference with the work performed on the system. Therefore, we write

$$W \equiv \mathrm{tr}[\hat{H}_{\mathrm{f}}\hat{\rho}_{\mathrm{f}}] - \mathrm{tr}[\hat{H}_{\mathrm{i}}\hat{\rho}_{\mathrm{i}}]. \tag{5.22}$$

By defining $\Delta F \equiv F_{\rm f} - F_{\rm i}$, we obtain

$$W \geq \Delta F, \tag{5.23}$$

which is, at least formally, the second law of thermodynamics for isothermal processes. The equality in (5.23) is achieved if and only if $\hat{\rho}_{\rm f} = \hat{\rho}_{\rm f,can}$.

We stress that we did not assume that the final state of the system is the canonical distribution. In fact, we cannot say that even the temperature is well-defined in the final state. The final free energy $F_{\rm f}$ is only formally defined by using the final Hamiltonian $\hat{H}_{\rm f}$ and the initial temperature T, which is a standard formulation in modern nonequilibrium statistical physics. Since the final state is arbitrary, inequality (5.23) can be applied to an arbitrary nonequilibrium processes in which only the initial state is in the canonical distribution; inequality (5.23) still holds even when the final state is far from equilibrium. We also stress that we did not assume that the system is large; inequality (5.23) can be applied even to small systems.

We note that the difference between the work and the free-energy change is given by the Kullback-Leibler divergence:

$$W - \Delta F = k_{\rm B} T S(\hat{\rho}_{\rm f} \| \hat{\rho}_{\rm f,can}), \tag{5.24}$$

which implies that the dissipation $W - \Delta F$ is given by the gap between the final state and the canonical distribution.

The purpose of this thesis is to consider small systems which is in contact with large heat bath(s). To explicitly take into account the effect of a heat bath, we divide the above system into two: small system **S** and large heat bath **B**. In this situation, we write the total Hamiltonian as

$$\hat{H}(\lambda, c) = \hat{H}^{\rm S}(\lambda) + \hat{H}^{\rm SB}(c) + \hat{H}^{\rm B}, \tag{5.25}$$

where $\hat{H}^{\rm S}(\lambda)$ is the Hamiltonian of **S**, $\hat{H}^{\rm SB}(c)$ is the interaction Hamiltonian between **S** and **B**, and $\hat{H}^{\rm B}(\lambda)$ is the Hamiltonian of **B**. We assume that **S** can be controlled through external parameters λ. In addition, we assume that the interaction between **S** and **B** can also be controlled by external parameters c. This assumption is not unrealistic: we can control the strength of the interaction by, for example, using an adiabatic wall on **S**. Moreover, in some special setups, we can use the dynamical decoupling to control the strength of the interaction. Let $\hat{H}_{\rm i} \equiv \hat{H}(\lambda(0), c(0))$, $\hat{H}_{\rm f} \equiv \hat{H}(\lambda(\tau), c(\tau))$, $\hat{H}_{\rm i}^{\rm S} \equiv \hat{H}^{\rm S}(\lambda(0))$, $\hat{H}_{\rm f}^{\rm S} \equiv \hat{H}(\lambda(\tau))$, $\hat{H}_{\rm i}^{\rm SB} \equiv \hat{H}^{\rm SB}(c(0))$, and $\hat{H}_{\rm f}^{\rm SB} \equiv \hat{H}^{\rm SB}(c(\tau))$. We define the partition functions and the Helmholtz free energies

$$Z_{\rm i}^{\rm S} \equiv {\rm tr}[e^{-\beta \hat{H}_{\rm i}^{\rm S}}], \quad Z_{\rm f}^{\rm S} \equiv {\rm tr}[e^{-\beta \hat{H}_{\rm f}^{\rm S}}], \tag{5.26}$$

and

$$F_{\rm i}^{\rm S} \equiv -k_{\rm B} T \ln Z_{\rm i}^{\rm S}, \quad F_{\rm f}^{\rm S} \equiv -k_{\rm B} T \ln Z_{\rm f}^{\rm S}. \tag{5.27}$$

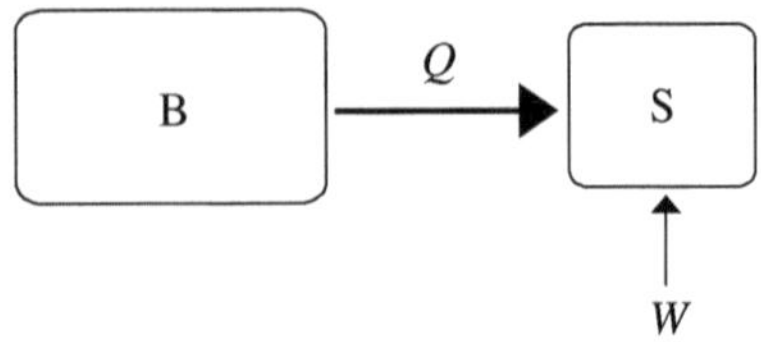

Fig. 5.1 Energy balance of the total system of **S** and **B** (reproduced from Ref. [4] with permission)

If the initial and final interactions are zero, i.e., $\hat{H}_{\mathrm{i}}^{\mathrm{SB}} = \hat{H}_{\mathrm{f}}^{\mathrm{SB}} = 0$ holds, then inequality (5.21) reduces to

$$W^{\mathrm{S}} \geq \Delta F^{\mathrm{S}}, \tag{5.28}$$

where

$$W^{\mathrm{S}} \equiv W \equiv \mathrm{tr}[\hat{H}_{\mathrm{f}}\hat{\rho}_{\mathrm{f}}] - \mathrm{tr}[\hat{H}_{\mathrm{i}}\hat{\rho}_{\mathrm{i}}] \tag{5.29}$$

is the work performed on **S**, and $\Delta F^{\mathrm{S}} \equiv F_{\mathrm{f}}^{\mathrm{S}} - F_{\mathrm{i}}^{\mathrm{S}}$ is the free-energy difference of **S**. We note that $W^{\mathrm{S}} = W$ holds because the total energy difference of **S** and **B** is the energy input through external parameter, which is the work (see also Fig. 5.1).

If the initial and final interactions are weak enough, inequality (5.28) holds approximately.

Inequality (5.28) is better than (5.23) as the microscopic foundation of the second law of thermodynamics (5.1), because the former takes into account the effect of a heat bath. In fact, in the original second law of thermodynamics, a macroscopic system obeys an isothermal process with a large heat bath, and both the initial and final states of the system are assumed to be in thermal equilibrium. If **S** and **B** are large enough and the system is relaxed to the thermal equilibrium in the final state, inequality (5.28) is expected to lead to the original second law of thermodynamics. However, we note that, inequality (5.28) is more general than the original second law of thermodynamics: in (5.28), we did not assume that **S** is large nor that the final state of **S** is in the canonical distribution.

We note that, if the initial and final interactions are zero nor weak, we can formally introduce the effective free energies [13] as

$$\tilde{F}_{\mathrm{i}}^{\mathrm{S}} \equiv F_{\mathrm{i}} - F^{\mathrm{B}}, \quad \tilde{F}_{\mathrm{f}}^{\mathrm{S}} \equiv F_{\mathrm{f}} - F^{\mathrm{B}}, \tag{5.30}$$

where $F^{\mathrm{B}} \equiv -k_{\mathrm{B}}T \ln \mathrm{tr}[e^{-\beta \hat{H}^{\mathrm{B}}}]$ is the free energy of **B**. Then inequality (5.23) trivially reduces to

$$W^{\mathrm{S}} \geq \Delta \tilde{F}^{\mathrm{S}}, \tag{5.31}$$

where $\Delta \tilde{F}^{\mathrm{S}} \equiv \tilde{F}_{\mathrm{f}}^{\mathrm{S}} - \tilde{F}_{\mathrm{i}}^{\mathrm{S}}$. However, in the following, we assume that $\hat{H}_{\mathrm{i}}^{\mathrm{SB}} = \hat{H}_{\mathrm{f}}^{\mathrm{SB}} = 0$ holds for simplicity.

5.3 General Situations with Multi-Heat Baths

We next consider a thermodynamic process of system **S** that can contact heat baths $\mathbf{B}_1$, $\mathbf{B}_2$, ..., $\mathbf{B}_n$, at respective temperatures $T_1 = (k_B\beta_1)$, $T_2 = (k_B\beta_2)$, ..., $T_n = (k_B\beta_n)$. We assume that the total of S and B_m's obeys a unitary evolution. The total Hamiltonian can be written as

$$\hat{H}(\lambda, \{c_m\}) = \hat{H}^{\mathrm{S}}(\lambda) + \sum_{m=1}^{n}(\hat{H}^{\mathrm{SB}_m}(c_m) + \hat{H}^{\mathrm{B}_m}), \tag{5.32}$$

where $\hat{H}^{\mathrm{S}}(\lambda)$ is the Hamiltonian of **S**, $\hat{H}^{\mathrm{SB}_m}(c_m)$ is the interaction Hamiltonian between **S** and $\mathbf{B}_m$, and $\hat{H}^{\mathrm{B}_m}$ is the Hamiltonian of $\mathbf{B}_m$. Here, λ describes controllable external parameters, and c_m describes external parameters to control the interaction between **S** and $\mathbf{B}_m$.

We consider a time evolution from 0 to τ, and assume that $\hat{H}^{\mathrm{SB}_m}(c_m(0)) = \hat{H}^{\mathrm{SB}_m}(c_m(\tau)) = 0$ holds for all m. We write $\hat{H}^{\mathrm{S}}(\lambda(0)) \equiv \hat{H}^{\mathrm{S}}_{\mathrm{i}}$, $\hat{H}^{\mathrm{S}}(\lambda(\tau)) \equiv \hat{H}^{\mathrm{S}}_{\mathrm{f}}$, $\hat{H}(\lambda(0), \{c_m(0)\}) \equiv \hat{H}_{\mathrm{i}}$, and $\hat{H}(\lambda(\tau), \{c_m(0)\}) \equiv \hat{H}_{\mathrm{f}}$.

We assume that the initial state of the total system is given by

$$\hat{\rho}_{\mathrm{i}} \equiv \hat{\rho}^{\mathrm{S}} \otimes \hat{\rho}^{\mathrm{B}_1}_{\mathrm{can}} \otimes \cdots \otimes \hat{\rho}^{\mathrm{B}_n}_{\mathrm{can}}, \tag{5.33}$$

where $\hat{\rho}^{\mathrm{S}}$ is an arbitrary initial state of **S**, and

$$\hat{\rho}^{\mathrm{B}_m}_{\mathrm{can}} \equiv \frac{e^{-\beta_m \hat{H}^{\mathrm{B}_m}}}{Z^{\mathrm{B}_m}}, \tag{5.34}$$

is the canonical distribution with $Z^{\mathrm{B}_m} = \mathrm{tr}[e^{-\beta \hat{H}^{\mathrm{B}_m}}]$. We write the free energies of the heat baths as $F^{\mathrm{B}_m} \equiv -k_B T_m \ln Z^{\mathrm{B}_m}$. We note that Eq. (5.33) is consistent with the assumption that $\hat{H}^{\mathrm{SB}_m}(c_m(0)) = \hat{H}^{\mathrm{SB}_m}(c_m(\tau)) = 0$ holds for all m.

The unitary evolution of the total system is given by

$$\hat{U} \equiv T\exp\left(-\mathrm{i}\int_0^{\tau} \hat{H}(\lambda(t), \{c_m(t)\})dt\right), \tag{5.35}$$

which leads to the final state

$$\hat{\rho}_{\mathrm{f}} \equiv \hat{U}\hat{\rho}_{\mathrm{i}}\hat{U}^{\dagger}. \tag{5.36}$$

We write $\hat{\rho}^{\mathrm{S}} \equiv \mathrm{tr}_{\mathrm{B}_1 \ldots \mathrm{B}_n}[\hat{\rho}_{\mathrm{f}}]$.

Due to the unitary invariance of the von Neumann entropy, we obtain

$$S(\hat{\rho}_{\mathrm{i}}) = S(\hat{\rho}_{\mathrm{f}}). \tag{5.37}$$

On the other hand, we have

$$S(\hat{\rho}_{\rm i}) = S(\hat{\rho}_{\rm i}^{\rm S}) + \sum_m \beta_m \left({\rm tr}[\hat{\rho}_{\rm i} \hat{H}^{{\rm B}_m}] - F^{{\rm B}_m} \right). \tag{5.38}$$

From Klein's inequality, we also have

$$S(\hat{\rho}_{\rm f}) \geq {\rm tr}[\hat{\rho}_{\rm f} \ln(\hat{\rho}_{\rm f}^{\rm S} \otimes \hat{\rho}_{\rm can}^{{\rm B}_1} \otimes \cdots \otimes \hat{\rho}_{\rm can}^{{\rm B}_n})] \tag{5.39}$$

$$= S(\hat{\rho}_{\rm f}^{\rm S}) + \sum_m \beta_m \left({\rm tr}[\hat{\rho}_{\rm f} \hat{H}^{{\rm B}_m}] - F^{{\rm B}_m} \right), \tag{5.40}$$

where we used $-{\rm tr}[\hat{\rho}_{\rm f} \ln \hat{\rho}_{\rm f}^{\rm S}] = S(\hat{\rho}_{\rm f}^{\rm S})$. Therefore we obtain

$$S(\hat{\rho}_{\rm f}^{\rm S}) - S(\hat{\rho}_{\rm i}^{\rm S}) \geq \sum_m \beta_m Q_m, \tag{5.41}$$

where

$$Q_m \equiv {\rm tr}[\hat{\rho}_{\rm i} \hat{H}^{{\rm B}_m}] - {\rm tr}[\hat{\rho}_{\rm f} \hat{H}^{{\rm B}_m}] \tag{5.42}$$

is the heat that is absorbed by system **S** from heat bath $\mathbf{B}_m$. Inequality (5.41) is the main result of this chapter. We stress that inequality (5.41) holds for arbitrary initial and final states of **S** (i.e., $\hat{\rho}_{\rm i}^{\rm S}$ and $\hat{\rho}_{\rm f}^{\rm S}$); in fact, we have only assumed that the initial distributions of the heat baths are in the canonical distribution. Inequality (5.41) can be regarded as a generalization of Clausius' inequality (5.4) to nonequilibrium initial and final distributions.

We consider inequality (5.41) for special cases.

Nonequilibrium steady state. We first consider a simple case in which system **S** is in contact with two heat baths $T_{\rm H} = (k_{\rm B}\beta_{\rm H})^{-1}$ and $T_{\rm L} = (k_{\rm B}\beta_{\rm L})^{-1}$ with $T_{\rm H} > T_{\rm L}$, and **S** is in a nonequilibrium steady state with a constant heat flow $Q_{\rm H} = -Q_{\rm L} \equiv Q$. Since **S** is in a steady state, we may assume $S(\hat{\rho}_{\rm i}^{\rm S}) = S(\hat{\rho}_{\rm f}^{\rm S})$. Therefore inequality (5.41) reduces to

$$(\beta_{\rm H} - \beta_{\rm L})Q \leq 0, \tag{5.43}$$

and therefore $Q \geq 0$, implying that the heat flows from the hot bath to the cold one.

Isothermal process. We next consider the case in which there is a single heat bath at temperature $T = (k_{\rm B}\beta)^{-1}$. In this case, inequality (5.41) reduces to

$$S(\hat{\rho}_{\rm f}^{\rm S}) - S(\hat{\rho}_{\rm i}^{\rm S}) \geq \beta Q. \tag{5.44}$$

We then assume that the initial state of **S** is the canonical distribution as

$$\hat{\rho}_{\rm i}^{\rm S} = \hat{\rho}_{\rm can,i}^{\rm S} \equiv \frac{e^{-\beta \hat{H}_{\rm i}^{\rm S}}}{Z_{\rm i}^{\rm S}}, \tag{5.45}$$

where $Z_{\rm i}^{\rm S} \equiv {\rm tr}[e^{-\beta \hat{H}_{\rm i}^{\rm S}}]$. We also introduce notations as

$$F_{\rm i}^{\rm S} \equiv -k_{\rm B} T \ln Z_{\rm i}^{\rm S}, \tag{5.46}$$

$$\rho_{\rm can,f}^{\rm S} \equiv \frac{e^{-\beta \hat{H}_{\rm f}^{\rm S}}}{Z_{\rm f}^{\rm S}}, \quad Z_{\rm f}^{\rm S} \equiv {\rm tr}[e^{-\beta \hat{H}_{\rm f}^{\rm S}}], \quad F_{\rm f}^{\rm S} \equiv -k_{\rm B} T \ln Z_{\rm f}^{\rm S}. \tag{5.47}$$

From Klein's inequality, we obtain

$$S(\hat{\rho}_{\rm f}^{\rm S}) - S(\hat{\rho}_{\rm i}^{\rm S}) \leq \beta(\Delta E^{\rm S} - \Delta F^{\rm S}), \tag{5.48}$$

where

$$\Delta E^{\rm S} \equiv {\rm tr}[\hat{H}_{\rm f}^{\rm S} \hat{\rho}_{\rm f}^{\rm S}] - {\rm tr}[\hat{H}_{\rm i}^{\rm S} \hat{\rho}_{\rm i}^{\rm S}] \tag{5.49}$$

is the internal-energy difference of **S**, and

$$\Delta F^{\rm S} \equiv F_{\rm f}^{\rm S} - F_{\rm i}^{\rm S}. \tag{5.50}$$

From inequalities (5.44) and (5.48), we have

$$\Delta E^{\rm S} - \Delta F^{\rm S} \geq Q. \tag{5.51}$$

On the other hand, the first law of thermodynamics holds as

$$\Delta E^{\rm S} = Q + W^{\rm S}, \tag{5.52}$$

where $W^{\rm S}$ is the work performed on the system, which is given by Eq. (5.29). Therefore we reproduce inequality (5.28).

With multi-heat baths. We next consider the case in which there are multi-heat baths. We again assume that the initial state of **S** is the canonical distribution at temperature $T = (k_{\rm B}\beta)^{-1}$. While T is arbitrary in general, we can assume that T is equal to T_m if **S** is initially in contact only with $\mathbf{B}_m$. By using inequality (5.48), inequality (5.41) leads to

$$\beta(\Delta E^{\rm S} - \Delta F^{\rm S}) \geq \sum_m \beta_m Q_m. \tag{5.53}$$

In the case of a thermodynamic cycle with $\Delta E^{\rm S} = \Delta F^{\rm S} = 0$, we obtain

$$\sum_m \beta_m Q_m \leq 0, \tag{5.54}$$

which is the Clausius' inequality (5.4). In particular, if there are two heat baths at temperatures $T_{\rm H} = (k_{\rm B}\beta_{\rm H})^{-1}$ and $T_{\rm L} = (k_{\rm B}\beta_{\rm L})^{-1}$ with $T_{\rm H} > T_{\rm L}$, we obtain

$$\frac{\mathrm{W}_{\mathrm{ext}}^{\mathrm{S}}}{\mathrm{Q}_{\mathrm{H}}} \leq 1 - \frac{T_{\mathrm{H}}}{T_{\mathrm{L}}}, \tag{5.55}$$

where $W_{\mathrm{ext}}^{\mathrm{S}} = Q_{\mathrm{H}} - Q_{\mathrm{L}}$ is the work that is extracted from the cycle. Inequality (5.55) implies the Carnot bound.

References

1. J. Kurchan, arXiv:cond-mat/0007360 (2000).
2. H. Tasaki, arXiv:cond-mat/0009244 (2000).
3. C. Jarzynski, D.K. Wójcik, Phys. Rev. Lett. **92**, 230602 (2004)
4. T. Sagawa, Prog. Theor. Phys. **127**, 1 (2012)
5. T. Sagawa, Second law-Like inequalities with quantum relative entropy: an introduction. In: Lectures on Quantum Computing, Thermodynamics and Statistical Physics, Kinki University Series on Quantum Computing (World Scientific, Hackensack, 2012) e-print: arXiv:1202.0983 (To appear).
6. S. Carnot, Réflexions sur la pussance motrice du feu et sur les machines propresà développer atte puissance, (Bachelier, 1824).
7. L. Tisza, P.M. Quay, Annal. Phys. **25**, 48 (1963)
8. E.H. Lieb, J. Yngvason, Phys. Rept. **310**, 1 (1999)
9. H.B. Callen, Thermodynamics and an Introduction to Thermostatistics, 2nd edn. (Wiley, New York, 1985).
10. H. Tasaki, Thermodynanmics–From a Modern Point of View (Baifu-kan, Tokyo, 2000) (in Japanese).
11. S. Sasa, Introduction to Thermodynamics (Kyoritsu, 2000), (in Japanese).
12. A. Shimizu, *Principles of Thermodynamics* (Univ, Tokyo Press, 2007). (in Japanese)
13. M. Campisi, P. Talkner, P. Hänggi, Phys. Rev. Lett. **102**, 210401 (2009)

Chapter 6
Second Law with Feedback Control

Abstract In this chapter, we derive a generalized second law with a quantum measurement and quantum feedback control [1, 2], which is the first main part of this thesis. Thermodynamics of feedback control has been studied for a long time in terms of Maxwell's demon [3–7]. In fact, the demon can be formulated as a feedback controller acting on thermodynamic systems. Recently, this topic has attracted numerous attentions based on the advancements in nonequilibrium statistical mechanics and quantum information theory [1, 2, 8–21]. In Sect. 6.1, we derive the lower bound of the entropy difference by feedback control. Based on it, in Sect. 6.2, we derive a generalized second law of thermodynamics with feedback control. In Sect. 6.3, we discuss a generalized Szilard engine [22] that achieves the upper bound of the generalized second law.

6.1 Entropy Inequality

We first derive the entropy balance of a quantum system that obeys a quantum measurement and quantum feedback control [23] in addition to unitary evolutions. Let $\hat{\rho}_{\rm i}$ be an arbitrary initial density operator of a finite-dimensional quantum system, which evolves as follows.

Step 1: Unitary evolution. From time 0 to t_1, the system undergoes unitary evolution $\hat{U}_{\rm i}$. At time t_1, the density operator is given by $\hat{\rho}_1 = \hat{U}_{\rm i}\hat{\rho}_{\rm i}\hat{U}_{\rm i}^\dagger$.

Step 2: Measurement. From time t_1 to t_2, we perform a quantum measurement on the system. We assume that the measurement is described by measurement operators $\{\hat{M}_k\}$ with k's being measurement outcomes, which leads to POVM

$$\hat{E}_k \equiv \hat{M}_k^\dagger \hat{M}_k. \tag{6.1}$$

T. Sagawa, *Thermodynamics of Information Processing in Small Systems*,
Springer Theses, DOI: 10.1007/978-4-431-54168-4_6,

We obtain each outcome k with probability

$$p_k = \mathrm{tr}(\hat{E}_k \hat{\rho}_1). \tag{6.2}$$

Here we assumed that every single measurement operator corresponds a single measurement outcome as Eq. (6.1). Let $\boldsymbol{p} \equiv \{p_k\}$. The post-measurement state corresponding to outcome k is given by

$$\hat{\rho}_2^{(k)} = \frac{1}{p_k} \hat{M}_k \hat{\rho}_1 \hat{M}_k^\dagger, \tag{6.3}$$

and the ensemble average is given by

$$\hat{\rho}_2 = \sum_k p_k \hat{\rho}_2^{(k)} = \sum_k \hat{M}_k \hat{\rho}_1 \hat{M}_k^\dagger. \tag{6.4}$$

Step 3: Feedback control. From t_2 to t_3, we perform feedback control; the corresponding unitary operator $\hat{U}_k$ depends on measurement outcome k. The post-feedback state corresponding to outcome k is given by

$$\hat{\rho}_3^{(k)} = \hat{U}_k \hat{\rho}_2^{(k)} \hat{U}_k^\dagger = \frac{1}{p_k} \hat{U}_k \hat{M}_k \hat{\rho}_1 \hat{M}_k^\dagger \hat{U}_k^\dagger, \tag{6.5}$$

and the ensemble average is given by

$$\hat{\rho}_3^{(k)} \equiv \sum_k p_k \hat{U}_k \hat{\rho}_2^{(k)} \hat{U}_k^\dagger = \sum_k \hat{U}_k \hat{M}_k \hat{\rho}_1 \hat{M}_k^\dagger \hat{U}_k^\dagger. \tag{6.6}$$

Step 4: Unitary evolution. After the feedback, from time t_3 to τ, the system evolves according to unitary operator $\hat{U}_\mathrm{f}$ which is independent of outcome k. The final state is $\hat{\rho}_\mathrm{f} = \hat{U}_\mathrm{f} \hat{\rho}_3 \hat{U}_\mathrm{f}^\dagger$.

The entire time evolution is then given by

$$\hat{\rho}_\mathrm{f} = \mathscr{E}(\hat{\rho}_\mathrm{i}) \equiv \sum_k \hat{U}_\mathrm{f} \hat{U}_k \hat{M}_k \hat{U}_\mathrm{i} \hat{\rho}_\mathrm{i} \hat{U}_\mathrm{i}^\dagger \hat{M}_k^\dagger \hat{U}_k^\dagger \hat{U}_\mathrm{f}^\dagger. \tag{6.7}$$

The difference in the von Neumann entropy S between the initial and final states can be bounded as follows:

$$\begin{aligned}
& S(\hat{\rho}_\mathrm{i}) - S(\hat{\rho}_\mathrm{f}) \\
& = S(\hat{\rho}_1) - S(\hat{\rho}_3) \\
& \leq S(\hat{\rho}_1) - \sum_k p_k S(\hat{\rho}_3^{(k)})
\end{aligned}$$

$$= S(\hat{\rho}_1) - \sum_k p_k S(\hat{\rho}_2^{(k)})$$

$$= S(\hat{\rho}_1) + \sum_k \mathrm{tr}\left(\sqrt{\hat{E}_k}\hat{\rho}_1\sqrt{\hat{E}_k}\ln\frac{\sqrt{\hat{E}_k}\hat{\rho}_1\sqrt{\hat{E}_k}}{p_k}\right)$$

$$= S(\hat{\rho}_1) + H(\boldsymbol{p}) + \sum_k \mathrm{tr}(\sqrt{\hat{E}_k}\hat{\rho}_1\sqrt{\hat{E}_k}\ln\sqrt{\hat{E}_k}\hat{\rho}_1\sqrt{\hat{E}_k}), \tag{6.8}$$

where $H(\boldsymbol{p}) \equiv -\sum_k p_k \ln p_k$ is the Shannon information obtained by the measurement. Note that in deriving the inequality (6.8), we used the convexity of the von Neumann entropy, i.e. $S(\sum_k p_k\hat{\rho}_3^{(k)}) \geq \sum_k p_k S(\hat{\rho}_3^{(k)})$. From the definition (4.83) of QC-mutual information [27–29], we obtain

$$S(\hat{\rho}_\mathrm{i}) - S(\hat{\rho}_\mathrm{f}) \leq I_\mathrm{QC}, \tag{6.9}$$

where the equality is achieved if and only if all of $\hat{\rho}_3^{(k)}$'s are the same. Intuitively speaking, this condition means that the feedback control is perfect, i.e., we used all the obtained information by feedback control.

We note that Nielsen et al. [10] have derived inequality $S(\hat{\rho}_\mathrm{i}) - S(\hat{\rho}_\mathrm{f}) \leq S(\hat{\rho}_\mathrm{i}, \mathcal{E})$, where $S(\hat{\rho}_\mathrm{i}, \mathcal{E})$ is the entropy exchange. The entropy exchange depends on the total process $\mathcal{E}$ including the feedback process. On the other hand, our inequality (6.9) is bounded by I_QC which does not depend on the feedback process, but only depends on the pre-measurement state and the POVM.

6.2 Generalized Second Laws

We now consider the energetics of feedback control on thermodynamic systems, in terms of the work, the heat, and the free energy. We consider a thermodynamic process of system **S** which can be in contact with heat baths $\mathbf{B}_1$, $\mathbf{B}_2$, ..., $\mathbf{B}_n$, at respective temperatures $T_1 = (k_\mathrm{B}\beta_1)$, $T_2 = (k_\mathrm{B}\beta_2)$, ..., $T_n = (k_\mathrm{B}\beta_n)$. In the following, we use the same notations as in Sect. 5.2.

We assume that the total system of **S** and heat baths $\mathbf{B}_m$ obeys a unitary evolution except for the process of a measurement. Apart from the measurement apparatus, the total Hamiltonian can be written as

$$\hat{H}(\lambda, \{c_m\}) = \hat{H}^\mathrm{S}(\lambda) + \sum_{m=1}^{n}(\hat{H}^{\mathrm{SB}_m}(c_m) + \hat{H}^{\mathrm{B}_m}). \tag{6.10}$$

For simplicity of notations, we write $\boldsymbol{\lambda} \equiv (\lambda, \{c_m\})$. We assume that the initial state of the total system is given by

$$\hat{\rho}_{\mathrm{i}} \equiv \hat{\rho}_{\mathrm{i}}^{\mathrm{S}} \otimes \hat{\rho}_{\mathrm{can}}^{\mathrm{B}_1} \otimes \cdots \otimes \hat{\rho}_{\mathrm{can}}^{\mathrm{B}_n}, \tag{6.11}$$

where $\hat{\rho}_{\mathrm{i}}^{\mathrm{S}}$ is an arbitrary initial state of **S**. The total density operator evolves as described in Step 1 to 5 in the previous chapter, which corresponds to the present setup as follows.

Step 1: From time 0 to t_1, the unitary operator is given by

$$\hat{U}_{\mathrm{i}} = T \exp\left(-\mathrm{i} \int_0^{t_1} \hat{H}(\boldsymbol{\lambda}(t))dt\right). \tag{6.12}$$

Step 2: From time t_1 to t_2, we perform a measurement. While we perform it only on **S**, the corresponding measurement operators can be extended to the total system, which are described by M_k's.

Step 3: From time t_2 to t_3, we perform feedback control in which the control protocol of $\boldsymbol{\lambda}$ depends on measurement outcome k as $\lambda(t;k)$. The unitary evolution is then given by

$$\hat{U}_k = T \exp\left(-\mathrm{i} \int \hat{H}(\boldsymbol{\lambda}(t;k))dt\right). \tag{6.13}$$

Step 4: From time t_3 to τ, the total system obeys the unitary evolution

$$\hat{U}_{\mathrm{f}} = T \exp\left(-\mathrm{i} \int_{t_3}^{\tau} \hat{H}(\boldsymbol{\lambda}(t))dt\right). \tag{6.14}$$

To transform inequality (6.9) to energetic inequalities, we can apply the same argument as in Sect. 5.2. First of all, inequality (5.37) is replaced by inequality (6.9) in the presence of feedback control. We then obtain [1, 2]

$$S(\hat{\rho}_{\mathrm{f}}^{\mathrm{S}}) - S(\hat{\rho}_{\mathrm{i}}^{\mathrm{S}}) \geq \sum_m \beta_m Q_m - I_{\mathrm{QC}}, \tag{6.15}$$

which is a generalization of inequality (5.41) to the situation in which the system is subject to feedback control. Inequality (6.15) is the main result in this chapter.

If there is a single heat bath and the initial distribution of **S** is a canonical distribution, (6.15) reduces to [1, 2]

$$W^{\mathrm{S}} \geq \Delta F^{\mathrm{S}} - k_{\mathrm{B}} T I_{\mathrm{QC}}, \tag{6.16}$$

where W^{S} is given by Eq. (5.29). Inequality (6.16) is a generalization of inequality (5.44) to feedback-controlled processes. By introducing notation $W_{\mathrm{ext}}^{\mathrm{S}} \equiv -W^{\mathrm{S}}$, inequality (6.16) can be rewritten as

$$W_{\mathrm{ext}}^{\mathrm{S}} \leq -\Delta F^{\mathrm{S}} + k_{\mathrm{B}} T I_{\mathrm{QC}}. \tag{6.17}$$

Inequality (6.17) implies that we can extract work greater than $-\Delta F^{\mathrm{S}}$ from a single heat bath with feedback control, but that we cannot extract work larger than $-\Delta F^{\mathrm{S}}+k_{\mathrm{B}}T I_{\mathrm{QC}}$. If $I_{\mathrm{QC}}=0$, inequality (6.17) reduces to (5.28). On the other hand, in the case of a classical and error-free measurement, inequality (6.17) becomes $W_{\mathrm{ext}} \leq -\Delta F^{\mathrm{S}}+k_{\mathrm{B}}TH(\boldsymbol{p})$.

The upper bound of inequality (6.17) can be achieved with the Szilard engine [24] in which $I_{\mathrm{QC}}=H(\boldsymbol{p})=\ln 2$, $W_{\mathrm{ext}}=\ln 2$, and $\Delta F^{\mathrm{S}}=0$ hold. In fact, in the case of the Szilard engine, the expansion is quasi-static and the post-feedback state is independent of the measurement outcomes. Moreover, as shown in Ref. [15], the upper bound of inequality (6.17) can be achieved for any quantum measurement satisfying Eq. (6.1). Some models that achieves the upper bound of (6.17) are discussed in Refs. [25, 26] for classical stochastic systems.

If there are multi-heat baths and the initial state of **S** is the canonical distribution at temperature $T=(k_{\mathrm{B}}\beta)^{-1}$, inequality (6.17) leads to the following inequality [1, 2]:

$$\beta(\Delta E^{\mathrm{S}}-\Delta F^{\mathrm{S}}) \geq \sum_{m} \beta_m Q_m - I_{\mathrm{QC}}, \tag{6.18}$$

which is a generalization of inequality (5.53). Inequality (6.18) represents the second law of thermodynamics with multi-heat baths in the presence of a feedback control, where the effect of the feedback control is described by the last term. For a thermodynamic cycle with $\Delta E^{\mathrm{S}}=0$, and $\Delta F^{\mathrm{S}}=0$, inequality (6.18) reduces to a generalized Clausius inequality

$$\sum_{m=1}^{n} \frac{Q_m}{T_m} \leq I_{\mathrm{QC}}. \tag{6.19}$$

In particular, in the case of a thermodynamic cycle with two heat baths with feedback control, we obtain

$$W_{\mathrm{ext}}^{\mathrm{S}} < \left(1-\frac{T_{\mathrm{L}}}{T_{\mathrm{H}}}\right) Q_{\mathrm{H}} + k_{\mathrm{B}} T_{\mathrm{L}} I_{\mathrm{QC}}, \tag{6.20}$$

which is a generalization of Carnot's bound (5.55). Inequality (6.20) implies that the upper bound for the efficiency of heat cycles becomes larger than that of the Carnot cycle with feedback control.

We can achieve the upper bound of (6.20) by performing a Szilard-type operation during an isothermal process of the one-molecule Carnot cycle. If we perform the measurement and feedback in the same scheme as the Szilard engine during the isothermal process at temperature T_{L}, we can extract the work of $W_{\mathrm{ext}}^{\mathrm{S}}=(1-T_{\mathrm{L}}/T_{\mathrm{H}})Q_{\mathrm{H}}+k_{\mathrm{B}}T_{\mathrm{L}}\ln 2$ in the total process. Moreover, if we perform the measurement and feedback in the same scheme as the Szilard engine during the isothermal process at temperature T_{H}, we can also obtain the same bound: $W_{\mathrm{ext}}^{\mathrm{S}}=(1-T_{\mathrm{L}}/T_{\mathrm{H}})(Q_{\mathrm{H}}-k_{\mathrm{B}}T_{\mathrm{H}}\ln 2)+k_{\mathrm{B}}T_{\mathrm{H}}\ln 2=(1-T_{\mathrm{L}}/T_{\mathrm{H}})Q_{\mathrm{H}}+k_{\mathrm{B}}T_{\mathrm{L}}\ln 2$.

6.3 Generalized Szilard Engines

As an example with a classical measurement, we discuss the Szilard engine with measurement errors [22], which achieves the upper bound of (6.17). The control protocol of the generalized Szilard engine is given by the following steps, which are described Fig. 6.1.

Step 1: Initial state. A single-molecular classical gas is in a box. The initial state of the gas is in thermal equilibrium with a single heat bath at temperature $T = (k_B\beta)^{-1}$.

Step 2: Insertion the barrier. We insert a barrier in the middle of the box, and divide it to two boxes with the same volume. Here, we do not know in which box the particle is. For simplicity of notations, we write "left" as "0" and "right" as "1." In other words, the position x of the particle is given by $x = 0$ or $x = 1$. We do not need any work during this process as proved in [16].

Step 3: Measurement. We measure the position of the particle. We assume that the measurement is equivalent to the binary symmetric channel with error rate ε discussed in Chap. 3. We write the measurement outcome as y (= 0 or 1). We note that $x = y$ holds for the original Szilard engine without error ($\varepsilon = 0$).

Step 4: Feedback. We next move the position of the barrier quasi-statically and isothermally. The protocol of moving the barrier depends on measurement outcome y. Let v_0 $(0 \leq v_0 \leq 1)$ and v_1 $(0 \leq v_0 \leq 1)$ be real numbers. We assume that, after we move the barrier, the ratio of the volumes of the boxes is assumed to be $v_0 : 1 - v_0$ for $y = 0$, or $1 - v_1 : v_1$ for $y = 1$. We note that, in the case of the original Szilard engine, $v_0 = v_1 = 1$ holds. In this process, we extract the work from the engine. The amounts of the work are given by $k_B T \ln 2v_0$ if $(x, y) = (0, 0)$, $k_B T \ln 2(1 - v_0)$ if $(x, y) = (0, 1)$, $k_B T \ln 2(1-v_1)$ if $(x, y) = (1, 0)$, and $k_B T \ln 2v_1$ if $(x, y) = (1, 1)$. The feedback protocol is characterized by v_0 and v_1.

Step 5: Removal of the barrier. We remove the barrier without any work. The engine then returns to the initial state.

From the total process, we extract the average work

$$W_{\mathrm{ext}}^{\mathrm{S}} = k_B T \left(\ln 2 + \frac{1-\varepsilon}{2} \ln v_0 + \frac{\varepsilon}{2} \ln(1 - v_0) + \frac{\varepsilon}{2} \ln(1 - v_1) + \frac{1-\varepsilon}{2} \ln v_1 \right). \tag{6.21}$$

We note that $\Delta F^{\mathrm{S}} = 0$ holds. We then maximize $W_{\mathrm{ext}}^{\mathrm{S}}$ under a given measurement error ε by changing v_0 and v_1. The maximum value of $W_{\mathrm{ext}}^{\mathrm{S}}$ is achieved when

$$v_0 = v_1 = 1 - \varepsilon, \tag{6.22}$$

with which the maximum work is given by

$$W_{\mathrm{ext}}^{\mathrm{S}} = k_B T (\ln 2 + \varepsilon \ln \varepsilon + (1 - \varepsilon) \ln(1 - \varepsilon)). \tag{6.23}$$

On the other hand, as discussed in Chap. 3, the mutual information of the binary symmetric channel is given by

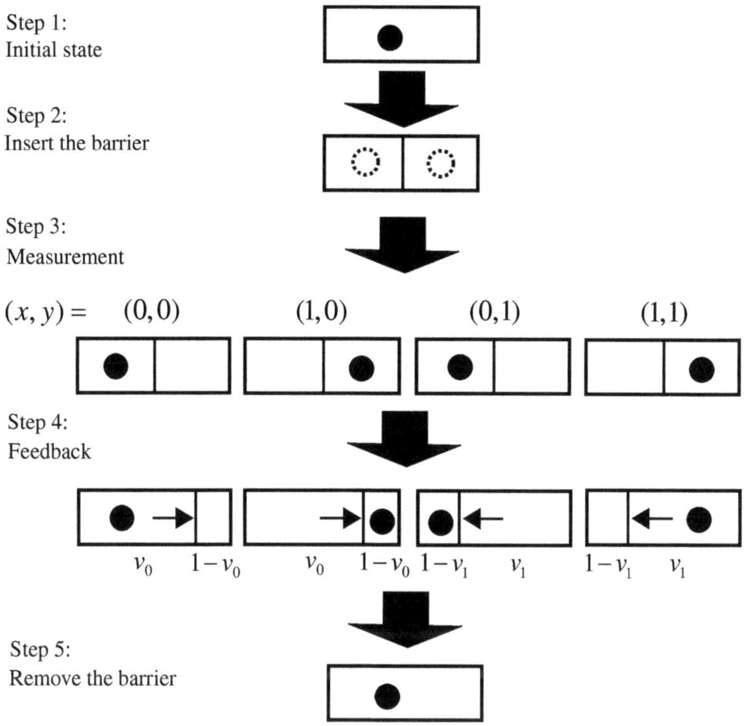

Fig. 6.1 The generalized szilard engine (reproduced from Ref. [22] with permission). ε is the measurement error rate, x denotes the states of the system, and y denote the measurement outcomes. The control protocol is determined by y

$$I = \ln 2 + \varepsilon \ln \varepsilon + (1 - \varepsilon) \ln(1 - \varepsilon). \tag{6.24}$$

Therefore we obtain

$$W_{\mathrm{ext}}^{\mathrm{S}} = k_{\mathrm{B}} T I, \tag{6.25}$$

which means that the generalized Szilard engine achieves the upper bound of the extractable work (6.17) for any amount of the mutual information.

In general, as shown in Ref. [15], the upper bound of inequality (6.17) can be achieved for any quantum measurement satisfying Eq. (6.1).

References

1. T. Sagawa, M. Ueda, Phys. Rev. Lett. **100**, 080403 (2008)
2. T. Sagawa, Prog. Theor. Phys. **127**, 1 (2012)
3. J.C. Maxwell, *Theory of Heat* (Appleton, London, 1871)

4. H.S. Leff, A.F. Rex (eds.), *Maxwell's Demon 2: Entropy, Classical and Quantum Information, Computing* (Princeton University Press, Princeton, 2003)
5. K. Maruyama, F. Nori, V. Vedral, Rev. Mod. Phys. **81**, 1 (2009)
6. O.J.E. Maroney, Information processing and thermodynamic entropy, in The Stanford Encyclopedia of Philosophy, Fall 2009 edn., ed. by E.N. Zalta (Fall, Chicago, 2009).
7. T. Sagawa, M Ueda, Information thermodynamics: maxwell's demon in nonequilibrium dynamics, in: Nonequilibrium Statistical Physics of Small Systems: Fluctuation Relations and Beyond. R. Klages, W. Just, C. and Jarzynski (Eds.),(Wiley, Weinheim, 2012) e-print : arXiv:1111.5769 (2011)(to appear).
8. S. Lloyd, Phys. Rev. A **39**, 5378 (1989)
9. S. Lloyd, Phys. Rev. A **56**, 3374 (1997)
10. M.A. Nielsen, C.M. Caves, B. Schumacher, H. Barnum, Proc. R. Soc. London A **454**, 277 (1998)
11. W.H. Zurek, arXiv:quant-ph/0301076 (2003).
12. T.D. Kieu, Phys. Rev. Lett. **93**, 140403 (2004)
13. A.E. Allahverdyan, R. Balian, ThM Nieuwenhuizen, J. Mod. Opt. **51**, 2703 (2004)
14. H.T. Quan, Y.D. Wang, Y-x. Liu, C. P. Sun, F. Nori. Phys. Rev. Lett. **97**, 180402 (2006)
15. K. Jacobs, Phys. Rev. A **80**, 012322 (2009)
16. S.W. Kim, T. Sagawa, S. De Liberato, M. Ueda, Phys. Rev. Lett. **106**, 070401 (2011)
17. Y. Morikuni, H. Tasaki, J. Stat. Phys. **143**, 1 (2011)
18. H. Dong, D.Z. Xu, C.Y. Cai, C.P. Sun, Phys. Rev. E **83**, 061108 (2011)
19. D.V. Averin, M. Möttönen, J.P. Pekola, Phys. Rev. B **84**, 245448 (2011)
20. S. Lahiri, S. Rana, A.M. Jayannavar, J. Phys. A: Math. Theor. **45**, 065002 (2012)
21. Y. Lu, G.L. Long, Phys. Rev. E **85**, 011125 (2012)
22. T. Sagawa, M. Ueda, Phys. Rev. E **85**, 021104 (2012)
23. H.M. Wiseman, G.J. Milburn, *Quantum Measurement and Control* (Cambridge University Press, Cambridge, 2010)
24. L. Szilard, Z. Phys. **53**, 840 (1929)
25. D. Abreu, U. Seifert, Europhys. Lett. **94**, 10001 (2011)
26. J.M. Horowitz, J.M.R. Parrondo, New J. Phys. **13**, 123019 (2011)
27. H.J. Groenewold, Int. J. Theor. Phys. **4**, 327 (1971)
28. M. Ozawa, J. Math. Phys. **27**, 759 (1986)
29. F. Buscemi, M. Hayashi, M. Horodecki, Phys. Rev. Lett. **100**, 210504 (2008)

Chapter 7
Thermodynamics of Memories

Abstract Maxwell's demon [1, 2] can be formulated as a feedback controller acting on thermodynamic systems, and has a memory that stores measurement outcomes.The fundamental energy cost needed for the memory during information processing has been a topic of numerous discussions [3–16], which is fundamentally related to the consistency between the demon and the second law of thermodynamics. In this chapter, we discuss the thermodynamic properties of the demon's memory based on recent advancements [5, 6, 17–25]. In particular, we identify the fundamental energy cost needed for the measurement and information erasure [23–25], which is the second main part of this thesis. Our result reduces to the celebrated Landauer principle [8] for special cases. We also discuss a new resolution of the paradox of Maxwell's demon based on our results. For simplicity, we consider the case in which there is a single heat bath in this chapter.

7.1 Formulation of Memory

We first formulate a memory **M** that stores measurement outcomes. We note that the "memory" may include the measurement apparatus that directly interacts with a measured system.

Let $\mathbf{H}^{\mathrm{M}}$ be the Hilbert space corresponding to **M**. We decompose $\mathbf{H}^{\mathrm{M}}$ into mutually orthogonal subspaces $\mathbf{H}_k^{\mathrm{M}}$ ($k = 0, 1, 2, \cdots, N$), where the k's describe the measurement outcomes. $\mathbf{H}^{\mathrm{M}}$ is written as the direct sum of $\mathbf{H}_k^{\mathrm{M}}$'s as

$$\mathbf{H}^{\mathrm{M}} = \bigoplus_k \mathbf{H}_k^{\mathrm{M}}. \tag{7.1}$$

Outcome "k" is stored in **M** if the support of the density operator of the memory belongs to $\mathbf{H}_k^{\mathrm{M}}$. We note that the classical outcomes (i.e., k's) can be distinguished each other when they are stored in **M**, because $\mathbf{H}_k^{\mathrm{M}}$'s are mutually orthogonal.

Springer Theses, DOI: 10.1007/978-4-431-54168-4_7, © Springer Japan 2013

Therefore, the assumption of the orthogonality is crucial for **M** to work as a memory that stores classical outcomes. We assume that **M** has a pre-fixed standard state, and that **M** is in the standard state before a measurement. We assume that $k = 0$ corresponds to the standard state of **M**.

The total Hamiltonian of **M**, denoted as $\hat{H}^{\mathrm{M}}$, can also be written as the direct sum of sub-Hamiltonians:

$$\hat{H}^{\mathrm{M}} \equiv \sum_k \hat{H}_k^{\mathrm{M}}, \tag{7.2}$$

where the support of $\hat{H}_k^{\mathrm{M}}$ is in $\mathbf{H}_k^{\mathrm{M}}$ for every k. Let the spectral decomposition of $\mathbf{H}_k^{\mathrm{M}}$ be $\hat{H}_k^{\mathrm{M}} \equiv \sum_i \varepsilon_{ki} |\varepsilon_{ki}\rangle\langle\varepsilon_{ki}|$, where $\{|\varepsilon_{ki}\rangle\}_i$ is an orthonormal basis set of $\mathbf{H}_k^{\mathrm{M}}$.

The conditional canonical distribution at temperature $T = (k_{\mathrm{B}}\beta)^{-1}$ under the condition of outcome "k" is given by

$$\hat{\rho}_{\mathrm{can},k}^{\mathrm{M}} \equiv \frac{e^{-\beta \hat{H}_k^{\mathrm{M}}}}{Z_k}, \tag{7.3}$$

where $Z_k^{\mathrm{M}} \equiv \mathrm{tr}[\exp(-\beta \hat{H}_k^{\mathrm{M}})]$. The corresponding Helmholtz free energy of **M** with "k" is given by

$$F_k^{\mathrm{M}} \equiv -k_{\mathrm{B}} T \ln Z_k^{\mathrm{M}}. \tag{7.4}$$

7.2 Erasure Process

We consider the following process for the information erasure in the presence of a single heat bath at temperature $T = (k_{\mathrm{B}}\beta)^{-1}$. The pre-erasure state means the post-measurement state, in which the memory stores the information of the measured system. In the pre-erasure state, **M** stores outcome "k" with probability p_k. We define $\boldsymbol{p} \equiv \{p_k\}$. The Shannon information corresponding to the pre-erasure state is given by

$$H(\boldsymbol{p}) \equiv -\sum_k p_k \ln p_k. \tag{7.5}$$

We assume that, before the information erasure, the state of **M** under the condition of "k" is in the canonical distribution $\hat{\rho}_k^{\mathrm{can}}$, and that the total pre-erasure state of **M** is given by

$$\hat{\rho}_{\mathrm{i}}^{\mathrm{M}} \equiv \sum_k p_k \hat{\rho}_{\mathrm{can},k}^{\mathrm{M}}. \tag{7.6}$$

We stress that memory **M** should be able to store an arbitrary probability distribution $\{p_k\}$, which is the most important necessary property for **M** to fulfill the function of a memory. For example, if the measurement is error-free, $\{p_k\}$ is determined only by the state of the measured system. In general, $\{p_k\}$ is independent of the structure of the memory, while F_k's are determined by the structure of the memory.

Let $\mathbf{H}^{\rm B}$ be the Hilbert space corresponding to the heat bath **B**. We assume that **B** is initially in the canonical distribution, which is given by $\hat{\rho}^{\rm B}_{\rm can} \equiv \exp(-\beta \hat{H}^{\rm B})/Z^{\rm B}$, where $\hat{H}^{\rm B}$ is the Hamiltonian of **B** and $Z^{\rm B} \equiv {\rm tr}[\exp(-\beta \hat{H}^{\rm B})]$ is the partition function. We also assume that the initial states of **M** and **B** are not correlated, and that the initial state of the total system is given by

$$\hat{\rho}^{\rm MB}_{\rm i} \equiv \hat{\rho}^{\rm M}_{\rm i} \otimes \hat{\rho}^{\rm B}_{\rm can}. \tag{7.7}$$

We consider the erasure process from $t = 0$ to $t = \tau$. During the erasure process, we change the Hamiltonian of **M** with a protocol which need to be independent of k. The total Hamiltonian at time t then is given by

$$\hat{H}^{\rm MB}(t) = \hat{H}^{\rm M}(t) + \hat{H}^{\rm int}(t) + \hat{H}^{\rm B}, \tag{7.8}$$

where $\hat{H}^{\rm int}(t)$ is the interaction Hamiltonian between **M** and **B**. We assume that $\hat{H}^{\rm M}(0) = \hat{H}^{\rm M}(\tau) = \hat{H}^{\rm M}$ and $\hat{H}^{\rm int}(0) = \hat{H}^{\rm int}(\tau) = 0$, which is consistent with Eq. (7.7). The time evolution of the total system from time 0 to τ is then given by the unitary operator $\hat{U} \equiv T\exp(-{\rm i}\int \hat{H}^{\rm MB}(t)dt)$, which gives the post-erasure state of the total systems

$$\hat{\rho}^{\rm MB}_{\rm f} = \hat{U}\hat{\rho}^{\rm MB}_{\rm i}\hat{U}^\dagger. \tag{7.9}$$

We define $\hat{\rho}^{\rm B}_{\rm f} \equiv {\rm tr}_{\rm M}[\hat{\rho}^{\rm MB}_{\rm f}]$ and $\hat{\rho}^{\rm M}_{\rm f} \equiv {\rm tr}_{\rm B}[\hat{\rho}^{\rm MB}_{\rm f}]$. After the information erasure, the state of **M** is in the standard state with unit probability. In other words, the support of $\hat{\rho}^{\rm MB}_{\rm f}$ is in $\mathbf{H}^{\rm M}_0 \otimes \mathbf{H}^{\rm B}$ with unit probability.

We now derive the minimal energy cost that is needed for the erasure process. From the general second law (5.41), we obtain

$$S(\hat{\rho}^{\rm M}_{\rm f}) - S(\hat{\rho}^{\rm M}_{\rm i}) \geq \beta Q^{\rm M}_{\rm eras}, \tag{7.10}$$

where

$$Q^{\rm M}_{\rm eras} \equiv {\rm tr}[\hat{H}^{\rm B}\hat{\rho}^{\rm B}_{\rm can}] - {\rm tr}[\hat{H}^{\rm B}\hat{\rho}^{\rm B}_{\rm f}] \tag{7.11}$$

is the heat that is absorbed in **M** during the erasure process. On the other hand, $S(\hat{\rho}^{\rm M}_{\rm i})$ can be decomposed as

$$S(\hat{\rho}^{\rm M}_{\rm i}) = \sum_k p_k S(\hat{\rho}^{\rm M}_{{\rm can},k}) + H(\boldsymbol{p}), \tag{7.12}$$

because $\hat{\rho}^{\rm M}_{{\rm can},k}$'s are mutually orthogonal. We then obtain

$$S(\hat{\rho}^{\rm M}_{\rm f}) - \sum_k p_k S(\hat{\rho}^{\rm M}_{{\rm can},k}) - H(\boldsymbol{p}) \geq \beta Q^{\rm M}_{\rm eras}. \tag{7.13}$$

We note that

$$S(\hat{\rho}^{\mathrm{M}}_{\mathrm{can},k}) = \beta(\mathrm{tr}[\hat{H}^{\mathrm{M}}_k \hat{\rho}^{\mathrm{M}}_{\mathrm{can},k}] - F^{\mathrm{M}}_k) \tag{7.14}$$

holds. On the other hand, from Klein's inequality, we have

$$S(\hat{\rho}^{\mathrm{M}}_{\mathrm{f}}) \leq -\mathrm{tr}[\hat{\rho}^{\mathrm{M}}_{\mathrm{f}} \ln \hat{\rho}^{\mathrm{M}}_{\mathrm{can},0}] = \beta(\mathrm{tr}[\hat{H}^{\mathrm{M}}_0 \hat{\rho}^{\mathrm{M}}_{\mathrm{f}}] - F^{\mathrm{M}}_0). \tag{7.15}$$

Therefore we obtain

$$-\Delta F^{\mathrm{M}}_{\mathrm{eras}} + \Delta E^{\mathrm{M}}_{\mathrm{eras}} - k_{\mathrm{B}} T H(\boldsymbol{p}) \geq Q^{\mathrm{M}}_{\mathrm{eras}}, \tag{7.16}$$

where

$$\Delta F^{\mathrm{M}}_{\mathrm{eras}} \equiv F^{\mathrm{M}}_0 - \sum_k p_k F^{\mathrm{M}}_k \tag{7.17}$$

is the difference of the averaged free energies of **M**, and

$$\Delta E^{\mathrm{M}}_{\mathrm{eras}} \equiv \mathrm{tr}[\hat{H}^{\mathrm{M}}_0 \hat{\rho}^{\mathrm{M}}_{\mathrm{f}}] - \sum_k p_k \mathrm{tr}[\hat{H}^{\mathrm{M}}_k \hat{\rho}^{\mathrm{M}}_{\mathrm{can},k}] \tag{7.18}$$

is the difference of the averaged internal energies of **M**. From the first law of thermodynamics,

$$\Delta E^{\mathrm{M}}_{\mathrm{eras}} = W^{\mathrm{M}}_{\mathrm{eras}} + Q^{\mathrm{M}}_{\mathrm{eras}} \tag{7.19}$$

holds, where $W^{\mathrm{M}}_{\mathrm{eras}}$ is the work that is needed for the erasure process. Therefore we obtain

$$W^{\mathrm{M}}_{\mathrm{eras}} \geq k_{\mathrm{B}} T H(\boldsymbol{p}) + \Delta F^{\mathrm{M}}_{\mathrm{eras}}, \tag{7.20}$$

which is the first main results in this chapter [23, 24]. We note that an inequality similar but not equivalent to (7.20) has also been derived in Ref. [21].

For the special case in which $F^{\mathrm{M}}_0 = F^{\mathrm{M}}_k$ for all k, $\Delta F^{\mathrm{M}} = 0$ holds. In this case, inequality (7.20) leads to

$$W^{\mathrm{M}}_{\mathrm{eras}} \geq k_{\mathrm{B}} T H(\boldsymbol{p}), \tag{7.21}$$

which is a general statement of Landauer's principle [8, 12, 15]. In this case, the minimal energy cost for the information erasure is proportional to the Shannon information of the measurement outcomes. If there are two outcomes "0" and "1" with $p_0 = p_1 = 1/2$ and $F_0 = f_1$, inequality (7.21) reduces to

$$W^{\mathrm{M}}_{\mathrm{eras}} \geq k_{\mathrm{B}} T \ln 2, \tag{7.22}$$

which is Landauer's principle discussed in Chap. 2

On the other hand, when $\Delta F^{\mathrm{M}} \neq 0$ holds, we can erase $H(\boldsymbol{p})$ of information with the work satisfying $W^{\mathrm{M}}_{\mathrm{eras}} < k_{\mathrm{B}} T H(\boldsymbol{p})$. For example, we can achieve $W^{\mathrm{M}}_{\mathrm{eras}} = 0$

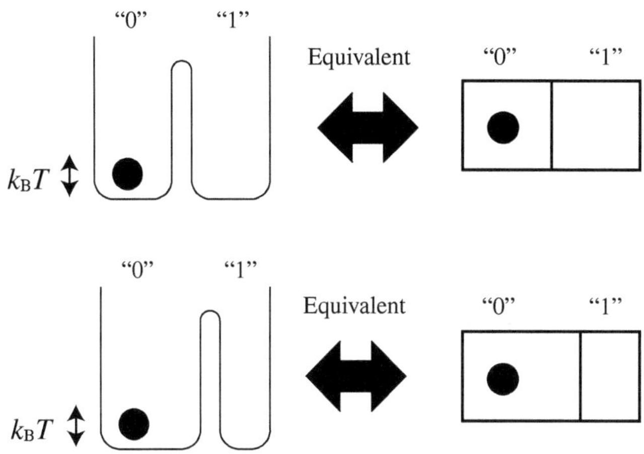

Fig. 7.1 Models of binary memories (reproduced from Ref. [25] with permission). A symmetric potential with $t = 1/2$ (*the upper row*) and an asymmetric potential with $t > 1/2$ (*the lower row*)

as discussed later. In this sense, there is no fundamental lower bound of the energy cost needed for the information erasure.

As an illustration, we consider a model of memory which can store a binary outcome "0" or "1" [23, 24]. Suppose that a Brownian particle is moving in a double-well potential (right column of Fig. 7.1) [8, 9], which is in contact with a single heat bath at temperature $T = (k_B\beta)^{-1}$. The particle is in the left well when the memory stores "0," and in the right well when the memory stores "1." We assume that the height of the barrier is sufficiently higher than both quantum and thermal fluctuations, so that the particle cannot exceed the barrier. With this assumption, the double-well potential is equivalent to two boxes (right column of Fig. 7.1). We note that the model illustrated in Fig. 7.1 is not for a measured system such as the Szilard engine [2]; rather it is only for the memory that stores the measurement outcome using the representation of a single-molecule gas.

Let $t : 7 - t$ $(0 < t < 1)$ be the ratio of the box. If $t = 1/2$, the memory is called symmetric. The memory can store arbitrary probability distribution of "0" and "1" (i.e., $p_0 \equiv p$ and $p_1 \equiv 1 - p$). We stress that p is independent of t; p is determined by the state of the measured system, while t characterizes the structure of the memory.

On the each box, the particle is assumed to be initially in thermal equilibrium under the condition of "0" or "1." The total initial state is in thermal equilibrium if and only if $t = p$, which we do not require in general. In the following, we assume that $p = 1/2$ for simplicity. We consider a quasi-static information erasure as shown in Fig. 7.2.

Step 1. In the initial state, the memory stores the measurement outcome "0" or "1."

Step 2. We then move the partition of the box (or the barrier of the potential) to the center. In this process, the average work is given by $(k_BT/2)[\ln 2t + \ln 2(1 - t)]$.

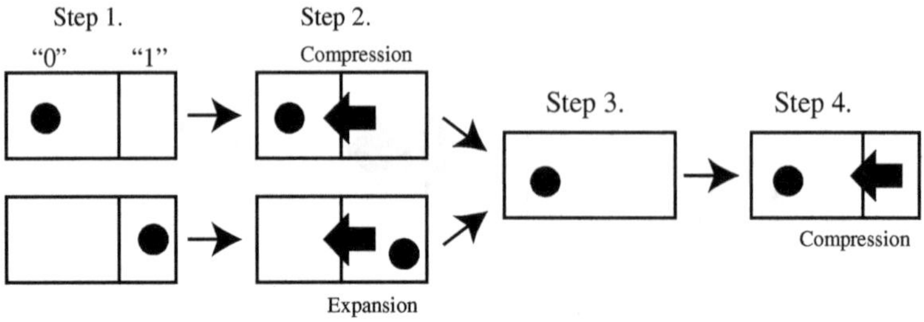

Fig. 7.2 A model of information erasure (reproduced from Ref. [25] with permission)

Step 3. We next remove the partition. This removal can be regarded as the free expansion of the gas, and therefore we do not need any work for the removal.

Step 4. We compress the box, and the memory returns to the standard state "0" with unit probability. The work of $-k_{\mathrm{B}}T\ln t$ is needed for this process.

The total work required for information erasure is given by

$$W^{\mathrm{M}}_{\mathrm{eras}} = k_{\mathrm{B}}T\ln 2 - (k_{\mathrm{B}}T/2)\ln(t/(1-t)). \tag{7.23}$$

If the memory is symmetric (i.e., $t = 1/2$), we have $W^{\mathrm{M}}_{\mathrm{eras}} = k_{\mathrm{B}}T\ln 2$ which achieve Landauer's bound [9]. On the other hand, we consider the case of, for example, $t = 4/5$. In this case, we have $W^{\mathrm{M}}_{\mathrm{eras}} = 0$, and therefore we do not need any work for the information erasure. In general, $W^{\mathrm{M}}_{\mathrm{eras}} < k_{\mathrm{B}}T\ln 2$ holds for $t > 1/2$. Therefore Landauer's principle for information erasure is valid for a symmetric double-well potential, but not for an asymmetric one. We note that the proof of $W^{\mathrm{M}}_{\mathrm{eras}} \geq k_{\mathrm{B}}T\ln 2$ using statistical mechanics in Ref. [12, 15] is valid only for the symmetric case. We also note that another asymmetric memory has also been discussed in Ref. [19].

7.3 Measurement Process

We next consider the measurement processes. Suppose that memory **M** performs a measurement on a measured system **S**, and stores outcome "k" with probability p_k. We assume that the memory is in contact with heat bath $\mathbf{B}_{\mathrm{M}}$ at temperature $T = (k_{\mathrm{B}}\beta)^{-1}$. On the other hand, during the measurement, measured system **S** adiabatically evolves or is in contact with a different heat bath, denoted as $\mathbf{B}_{\mathrm{S}}$, which is different from $\mathbf{B}_{\mathrm{M}}$. The latter assumption corresponds to the condition that the thermal noises on **M** and **S** are independent. For example, two colloidal particles, whose Langevin noises are independent, may satisfy this condition, even when they are actually in the same water. The total Hamiltonian is then given by

$$\hat{H}^{\mathrm{tot}}(t) = \hat{H}^{\mathrm{M}}(t) + \hat{H}^{\mathrm{MB_M}}(t) + \hat{H}^{\mathrm{B_M}} + \hat{H}^{\mathrm{S}}(t) + \hat{H}^{\mathrm{SB_S}}(t) + \hat{H}^{\mathrm{B_S}} + \hat{H}^{\mathrm{MS}}(t), \tag{7.24}$$

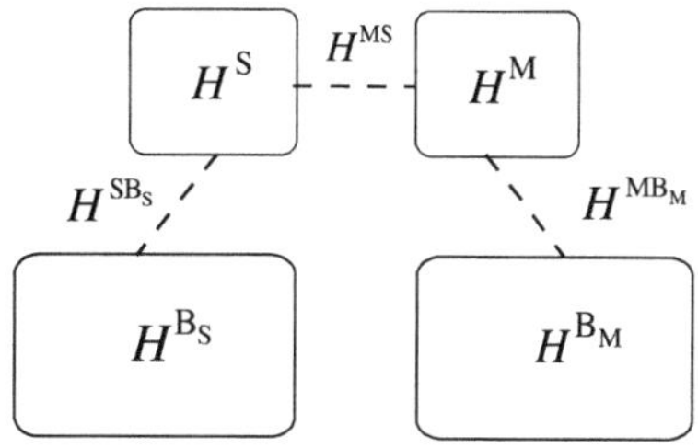

Fig. 7.3 A schematics of the interactions of the total system (reproduced from Ref. [25] with permission)

where $\hat{H}^{\mathrm{MS}}(t)$ describes the interaction between **M** and **S** for the measurement (see also Fig. 7.3). We assume that $\hat{H}^{\mathrm{M}}(0) = \hat{H}^{\mathrm{M}}(\tau) = \hat{H}^{\mathrm{M}}$ and $\hat{H}^{\mathrm{MB_M}}(0) = \hat{H}^{\mathrm{MB_M}}(\tau) = \hat{H}^{\mathrm{MS}}(0) = \hat{H}^{\mathrm{MS}}(\tau) = 0$ hold. We consider the following measurement process.

Step 1: Initial state. The initial state of **M** is in the standard state "0"; we assume that the initial state of **M** is the conditional canonical distribution under the condition that the support of the density operator is in $\mathbf{H}_0^{\mathrm{M}}$. Then the initial density operator of the total system is given by

$$\hat{\rho}_{\mathrm{i}}^{\mathrm{tot}} = \hat{\rho}_{\mathrm{can},0}^{\mathrm{M}} \otimes \hat{\rho}_{\mathrm{can}}^{\mathrm{B_M}} \otimes \hat{\rho}^{\mathrm{SB_S}}, \tag{7.25}$$

where $\hat{\rho}_{\mathrm{can}}^{\mathrm{B_M}}$ is the initial canonical distribution of $\mathbf{B}_{\mathrm{M}}$, and $\hat{\rho}^{\mathrm{SB_S}}$ is the initial density operator of $\mathbf{S} + \mathbf{B}_S$. We stress that, in this chapter, we do not put any assumption on $\hat{\rho}^{\mathrm{SB_S}}$.

Step 2: Unitary evolution. The total system evolves unitarily due to the Hamiltonian (7.24). We write

$$\hat{U}_{\mathrm{i}} \equiv T \exp\left(-\mathrm{i}\int_0^{t_1} \hat{H}^{\mathrm{tot}} dt\right). \tag{7.26}$$

By this interaction, memory **M** becomes entangled with measured system **S**. After this interaction, the total density operator is given by $\hat{U}_{\mathrm{i}}\hat{\rho}_{\mathrm{i}}^{\mathrm{tot}}\hat{U}_{\mathrm{i}}^{\dagger}$.

Step 3: Projection. The state of **M** is projected onto the subspace corresponding to the measurement outcome "k." This process is described by the projection operator on $\mathbf{H}_k^{\mathrm{M}}$ as

$$\hat{P}_k^{\mathrm{M}} \equiv \sum_i |\varepsilon_{ki}\rangle\langle\varepsilon_{ki}| \otimes \hat{I}^{\mathrm{B_M SB_S}}, \tag{7.27}$$

where $\hat{I}^{B_M S B_S}$ is the identity operator on $\mathbf{B}_{\mathrm{M}} + \mathbf{S} + \mathbf{B}_{\mathrm{S}}$. We note that, for the case of classical measurements, we do not need this projection process. Immediately after the measurement, the total density operator is given by

$$\hat{\rho}_{\rm f}^{\rm tot} = \sum_k \hat{P}_k^{\rm M} \hat{U}_{\rm i} \hat{\rho}_{\rm i}^{\rm tot} \hat{U}_{\rm i}^{\dagger} \hat{P}_k^{\rm M}. \tag{7.28}$$

We assume that the post-measurement state is given by

$$\hat{\rho}_{\rm f}^{\rm tot} = \sum_k \hat{M}_k \hat{\rho}_{\rm i}^{\rm SB_S} \hat{M}_k^{\dagger} \otimes \hat{\rho}_k^{\rm MB_M}, \tag{7.29}$$

where $\hat{M}_k$'s are the measurement operators, and $\hat{\rho}_k^{\rm MB}$'s are the density operators of **M** and $\mathbf{B}_M$ that are mutually orthogonal. Assumption (7.29) is equivalent to the assumption that any element of the POVM is given by a single measurement operator:

$$\hat{E}_k \equiv \hat{M}_k^{\dagger} \hat{M}_k. \tag{7.30}$$

We note that the probability of obtaining outcome "k" is given by

$$p_k = {\rm tr}(\hat{E}_k \hat{\rho}_{\rm i}^{\rm S}). \tag{7.31}$$

Let $H(\boldsymbol{p}) \equiv -\sum_k p_k \ln p_k$ be the Shannon information. We note that the following results in this chapter can be applied to both quantum and classical systems that satisfy Eq. (7.29).

We define the change in the averaged free energy due to the measurement as

$$\Delta F_{\rm meas}^{\rm M} \equiv \sum_k p_k F_k^{\rm M} - F_0^{\rm M}. \tag{7.32}$$

We note that $\Delta F_{\rm meas}^{\rm M} = -\Delta F_{\rm eras}^{\rm M}$ holds. On the other hand, the ensemble average of work performed on **M** during the measurement as

$$W_{\rm meas}^{\rm M} \equiv \sum_k p_k [{\rm tr}(\hat{\rho}_k^{\rm MB} \hat{H}_k^{\rm M}) + {\rm tr}(\hat{\rho}_k^{\rm MB} \hat{H}^{\rm B_M})] - [{\rm tr}(\hat{\rho}_{\rm can,0}^{\rm M} \hat{H}_0^{\rm M}) + {\rm tr}(\hat{\rho}_{\rm can}^{\rm B} \hat{H}^{\rm B_M})], \tag{7.33}$$

where we counted the energy flow between **S** and **M** through $\hat{H}^{\rm SM}$ as the work. In other words, $W_{\rm meas}^{\rm M}$ can be divided into the work through the change in the external parameter of **M** and that through the energy exchange between **S** and **M** during the measurement. We note that the definition of $W_{\rm meas}^{\rm M}$ is consistent with the definition of $W^{\rm S}$ in the previous chapter.

The second law of thermodynamics for the measurement process identifies the lower bound of $W_{\rm meas}^{\rm M}$:

$$W_{\rm meas}^{\rm M} \geq -k_{\rm B} T H(\boldsymbol{p}) + \Delta F_{\rm meas}^{\rm M} + k_{\rm B} T I_{\rm QC}, \tag{7.34}$$

where $I_{\rm QC}$ is the QC-mutual information corresponding to POVM $\{E_k\}$. We will prove this inequality later. Inequality (7.34) is the second main result in this chapter [23, 24], which gives the fundamental thermodynamic energy cost for measurement, regardless of the state of the measured system **S**. The right-hand side of (7.34) is an increasing function for a given value of H; the more effective information is obtained by the measurement, the more work is needed for the measurement.

For the special case that the measurement is error-free and classical (i.e., $I_{\rm QC} = H(\boldsymbol{p})$) and $\Delta F^{\rm M} = 0$ holds, inequality (7.34) reduces to

$$W^{\rm M}_{\rm meas} \geq 0, \tag{7.35}$$

which means is that there is no fundamental energy cost for measurement in this case [9].

We now prove inequality (7.34) under the assumption of (7.29). Since the von Neumann entropy is invariant under unitary evolutions and increases under projections, we have

$$S(\hat{\rho}^{\rm tot}_{\rm i}) \leq S(\hat{\rho}^{\rm tot}_{\rm f}). \tag{7.36}$$

On the other hand, from the assumption that $\hat{\rho}^{\rm MB}_k$'s are mutually orthogonal, we obtain

$$\begin{aligned} S(\hat{\rho}^{\rm tot}_{\rm f}) &= H(\boldsymbol{p}) + \sum_k p_k S\left(\hat{M}_k \hat{\rho}^{\rm SB_S}_{\rm i} \hat{M}^\dagger_k \otimes \hat{\rho}^{\rm MB_M}_k / p_k\right) \\ &= H(\boldsymbol{p}) + \sum_k p_k \left[S\left(\hat{M}_k \hat{\rho}^{\rm SB_S}_{\rm i} \hat{M}^\dagger_k / p_k\right) + S(\hat{\rho}^{\rm MB_M}_k)\right] \\ &= H(\boldsymbol{p}) + \sum_k p_k \left[S(\sqrt{\hat{E}_k} \hat{\rho}^{\rm SB_S}_{\rm i} \sqrt{\hat{E}_k} / p_k) + S(\hat{\rho}^{\rm MB_M}_k)\right]. \end{aligned} \tag{7.37}$$

From the definition of the QC-mutual information content $I_{\rm QC}$, we obtain

$$\sum_k p_k S(\hat{\rho}^{\rm MB_M}_k) - S(\hat{\rho}^{\rm M}_{\rm can,0}) - S(\hat{\rho}^{\rm B_M}_{\rm can}) \geq I_{\rm QC} - H(\boldsymbol{p}). \tag{7.38}$$

By using Klein's inequality, we have

$$-\sum_k p_k {\rm tr}[\hat{\rho}^{\rm MB_M}_k \ln \hat{\rho}^{\rm M}_{{\rm can},k} \otimes \hat{\rho}^{\rm B_M}_{\rm can}] - S(\hat{\rho}^{\rm M}_{\rm can,0}) - S(\hat{\rho}^{\rm B_M}_{\rm can}) \geq I_{\rm QC} - H(\boldsymbol{p}). \tag{7.39}$$

From the definition of the work, we finally obtain $-\Delta F^{\rm M}_{\rm meas} + W^{\rm M}_{\rm meas} \geq k_{\rm B} T (I_{\rm QC} - H(\boldsymbol{p}))$, which is inequality (7.34).

We next consider a model of quasi-static measurement process at temperature T [23, 24]. We consider the model of the classical binary memory shown in Fig. 7.1. In the initial state, the memory is in the standard state "0." We assume that the measurement is error-free. If the state of the measured system is given by "0," then the state of the memory does not change as show in the upper row of Fig. 7.4. If the

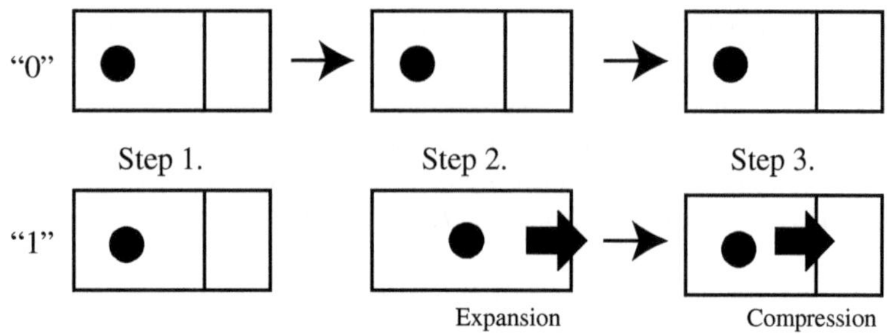

Fig. 7.4 A model of measurement (reproduced from Ref. [25] with permission)

measured state is in "1," the memory evolves as follows (see also the lower row of Fig. 7.4).

Step 1. The particle is in the left box corresponding to the standard state "0."

Step 2. The left box of the memory expands to the right. We need $-k_B T \ln(1/t)$ of work for this process.

Step 3. The box next compresses from the left until the volume of the right box returns to the initial volume, for which we need $k_B T \ln(1/(1-t))$ of work.

Therefore the total work for "1" is given by $k_B T \ln(t/(1-t))$. By averaging the work over the measurement outcomes "0" and "1," we find that

$$W^{\mathrm{M}}_{\mathrm{meas}} = (k_B T/2) \ln(t/(1-t)) \tag{7.40}$$

is required for the measurement on average.

7.4 Reconciliation with Maxwell's Demon

We next discuss the resolution of the paradox of Maxwell's demon [23, 24]. First of all, we sum up inequalities (7.34) and (7.20), and obtain

$$W^{\mathrm{M}}_{\mathrm{meas}} + W^{\mathrm{M}}_{\mathrm{eras}} \geq k_B T I_{\mathrm{QC}}, \tag{7.41}$$

which implies that there is a trade-off relation between the energy costs needed for the measurement and for the erasure. If the work needed for the information erasure is negative, the work needed for the measurement must be positive, and vice versa. Although there is no fundamental lower bound of the work for the measurement or only for the erasure, there exists a fundamental lower bound on their sum. We note that we have adopted the commonly used definitions for the measurement and the erasure [3]. Under the fixed condition for them, we can still change the ratio of $W^{\mathrm{M}}_{\mathrm{meas}}$ and $W^{\mathrm{M}}_{\mathrm{eras}}$ by changing the physical structure of the memory, which is described by $F^{\mathrm{M}}_{\mathrm{meas}} = -F^{\mathrm{M}}_{\mathrm{eras}}$.

We stress that the lower bound on the total energy cost does not depend on the Shannon information nor on the free-energy difference, but only on the QC-mutual information obtained by the measurement. This fact implies that the origin of the fundamental energy cost is not the randomness of the measurement outcomes which is described by the Shannon information; but is the correlation between the measured system and the memory which is described by the (QC-)mutual information.

We can illustrate the trade-off relation (7.41) by the model of the binary memory that has been discussed in the foregoing sections. By summing up the erasure cost (7.23) and the measurement cost (7.40), we find that the total work is given by

$$W^{\mathrm{M}}_{\mathrm{meas}} + W^{\mathrm{M}}_{\mathrm{eras}} = k_{\mathrm{B}} T \ln 2, \tag{7.42}$$

with which the equality in (7.41) is achieved. In fact, $I_{\mathrm{QC}} = H = \ln 2$ holds in our model.

We now discuss the consistency between the demon and the second law of thermodynamics. The fundamental upper bound of the work that can be extracted by the demon has been identified in Chap. 6. In particular, if the free-energy difference ΔF^{S} of the controlled system **S** is zero, the work that can be extracted by the demon is given by

$$W^{\mathrm{S}}_{\mathrm{ext}} \leq k_{\mathrm{B}} T I_{\mathrm{QC}}. \tag{7.43}$$

By summing up inequalities (7.41) and (7.43), we obtain

$$W^{\mathrm{SM}}_{\mathrm{ext}} \equiv W^{\mathrm{S}}_{\mathrm{ext}} - W^{\mathrm{M}}_{\mathrm{meas}} - W^{\mathrm{M}}_{\mathrm{eras}} \leq 0, \tag{7.44}$$

which implies that the work that can be extracted from the total system of **S** and **M** cannot be positive. This is consistent with Kelvin's principle (5.2) for the cycle of the total system. Therefore, the conventional second law of thermodynamics is satisfied for the total system. We note that the foregoing argument is valid under the assumption that any element of the POVM corresponds to a single measurement operator as $\hat{E}_k \equiv \hat{M}_k^{\dagger} \hat{M}_k$. To discuss the cases in which this assumption is not satisfied is a future challenge.

As discussed in Chap. 2, Brillouin argued that, based on a specific model, a positive work is needed for the measurement, which must be larger than the excess work extracted by Maxwell's demon [7]. After that, Bennett proposed a model that can perform measurement without any positive work. Moreover, he argued that, on the based on a specific model and Landauer's principle, we always need a positive work to erase the information stored in the memory [9], which has been widely accepted as the resolution of Maxwell's demon. Here, we have constructed a model with which we do not need any positive work for the information erasure. Moreover, we have derived inequalities (7.41) and (7.44), which enables us to finally reconcile Maxwell's demon with the second law of thermodynamics; what reconciles the demon with the second law is the total work of the measurement and erasure, which compensates for the excess work of $k_{\mathrm{B}} T I_{\mathrm{QC}}$ that can be extracted by the demon.

We note that $k_B T I_{QC}$ of work extracted by the demon can be still useful. By using feedback control, we can get the controlled system to obtain the free energy or the work even when there is no direct energy flow between the controller and the controlled system. We stress that, without feedback control, we need the direct energy input to the system in order to get it to obtain the free energy or the work.

7.5 Second Law of Information Thermodynamics

The conventional second law of thermodynamics cannot be applied to information processing straightforwardly. On the other hand, our main inequalities (6.17), (7.34), and (7.20) are the generalizations of the second law of thermodynamics for information processing processes. In fact, in the limit of $H \to 0$ and $I_{QC} \to 0$, all of inequalities (6.17), (7.34), and (7.20) reduce to the conventional second law of thermodynamics.

In the inequalities, information content (such as I_{QC} and H) and thermodynamic quantities (such as W and ΔF) are treated on an equal footing. Therefore, they constitute the second law of "information thermodynamics," which is the generalization of thermodynamics to information processing processes.

References

1. J.C. Maxwell, *Theory of Heat* (Appleton, London, 1871)
2. L. Szilard, Z. Phys. **53**, 840 (1929)
3. H.S. Leff, A.F. Rex (eds.), *Maxwell's Demon 2: Entropy, Classical and Quantum Information, Computing* (Princeton University Press, New Jersey, 2003)
4. K. Maruyama, F. Nori, V. Vedral, Rev. Mod. Phys. **81**, 1 (2009)
5. O.J.E. Maroney, in Information Processing and Thermodynamic Entropy, ed. by Edward N. Zalta. The Stanford Encyclopedia of Philosophy, Fall 2009 edn. (Institute of Physics Publishing, Philadelphia, 2009).
6. T. Sagawa, M. Ueda, Information Thermodynamics: Maxwell's Demon in Nonequilibrium dynamics, arXiv:1111.5769 (2011); To appear in R. Klages, W. Just, C, Jarzynski (eds.) Nonequilibrium Statistical Physics of Small Systems: Fluctuation Relations and Beyond (Wiley-VCH, Weinheim, 2012).
7. L. Brillouin, J. Appl. Phys. **22**, 334 (1951)
8. R. Landauer, IBM J. Res. Dev. **5**, 183 (1961)
9. C.H. Bennett, Int. J. Theor. Phys. **21**, 905 (1982)
10. W.H. Zurek, Nature **341**, 119 (1989)
11. W.H. Zurek, Phys. Rev. A **40**, 4731 (1989)
12. K. Shizume, Phys. Rev. E **52**, 3495 (1995)
13. R. Landauer, Science **272**, 1914 (1996)
14. H. Matsueda, E. Goto, K-F. Loe, RIMS Kôkyûroku **1013**, 187 (1997)
15. B. Piechocinska, Phys. Rev. A **61**, 062314 (2000)
16. C.H. Bennett, Stud. Hist. Phil. Mod. Phys. **34**, 501 (2003)
17. A.E. Allahverdyan, T.M. Nieuwenhuizen, Phys. Rev. E **64**, 0561171 (2001)
18. C. Horhammer, H. Buttner, J. Stat. Phys. **133**, 1161 (2008)

19. M.M. Barkeshli, arXiv:cond-mat/0504323 (2005).
20. J.D. Norton, Stud. Hist. Phil. Mod. Phys. **36**, 375 (2005)
21. O.J.E. Maroney, Phys. Rev. E **79**, 031105 (2009)
22. S. Turgut, Phys. Rev. E **79**, 041102 (2009)
23. T. Sagawa, M. Ueda, Phys. Rev. Lett. **102**, 250602 (2009)
24. T. Sagawa, M. Ueda, Phys. Rev. Lett. 106, 189901(E) (2011).
25. T. Sagawa, Prog. Theor. Phys. **127**, 1 (2012)

Chapter 8
Stochastic Thermodynamics

Abstract In this chapter, we review thermodynamics of classical stochastic systems, which has been referred to as stochastic thermodynamics [1–3]. In particular, we discuss nonequilibrium equalities such as the fluctuation theorem and the Jarzynski equality, which lead to the second law of thermodynamics as a corollary. These equalities have been theoretically derived in various setups [4–20] and experimentally verified by using small thermodynamic systems such as macromolecules [21, 22]. First, we formulate the general statement of the detailed fluctuation theorem and discuss its consequences. Second, we derive the detailed fluctuation theorem in Hamiltonian systems and Markovian stochastic systems.

8.1 Dynamics

We consider a classical stochastic system **S** that is in contact with heat baths $\mathbf{B}_1$, $\mathbf{B}_2$, ..., $\mathbf{B}_m$ at respective temperatures $T_1 = (k_B\beta_1)^{-1}$, $T_2 = (k_B\beta_2)^{-1}$, ..., $T_m = (k_B\beta_n)^{-1}$. Let x be the phase-space point of system **S** and λ be a set of external parameters such as the volume of a gas or the frequency of an optical tweezers. We control the system from time 0 to τ with control protocol $\lambda(t)$. Let $x(t)$ be the trajectory of the system.

To formulate the stochastic dynamics, we discretize the time interval $[0, \tau]$ by dividing it into N small intervals with width $\Delta t \equiv \tau/N$. The original continuous-time dynamics is recovered by taking $N \to \infty$ or equivalently $\Delta t \to 0$. Let $t = n\Delta t$ and $x_n \equiv x(n\Delta t)$. We refer to "time t" as "time $t_n \equiv n\Delta t$." Trajectory $\{x(t')\}_{t' \in [0,t]}$ corresponds to $X_n \equiv (x_0, x_1, \ldots, x_n)$.

Control protocol $\lambda(t)$ can also be discretized. Let λ_n be the value of λ between time $t_n = n\Delta t$ to $t_{n+1} = (n+1)\Delta t$, which is assumed to be constant during this time interval (see Fig. 8.1). We write the trajectory of λ from time 0 to t_n as $\Lambda_n \equiv (\lambda_0, \lambda_1, \ldots, \lambda_{n-1})$. Let $\lambda_{\rm int}$ be the value of parameter λ before time 0, which is not necessarily equal to λ_0 because we can switch the parameter at time 0. We also

T. Sagawa, *Thermodynamics of Information Processing in Small Systems*,
Springer Theses, DOI: 10.1007/978-4-431-54168-4_8, © Springer Japan 2013

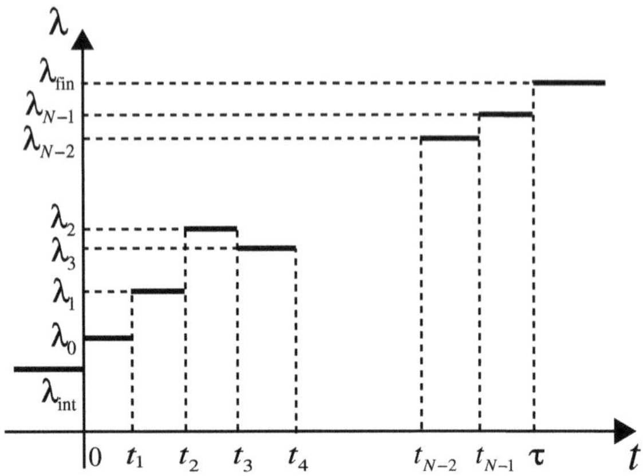

Fig. 8.1 Discretization of control protocol $\lambda(t)$ (reproduced from Ref. [23] with permission)

write the value of λ after time $t_N \equiv \tau$ as λ_{fin}, which is also not necessarily equal to λ_N, because we can switch the parameter at time t_N (see also Fig. 8.1).

Let $P_n[x_n]$ be the probability distribution of x at time t_n. In particular, $P_0[x_0]$ is the initial distribution of x. The initial distribution can be chosen as the stationary distribution under external parameters λ_{int}, denoted as $P_{\text{s}}[x_0|\lambda_{\text{int}}]$, which means $P_0[x_0] = P_{\text{s}}[x_0|\lambda_{\text{int}}]$. We note that $P_{\text{s}}[x_0|\lambda_{\text{int}}]$ is not necessarily a canonical distribution; it can be a nonequilibrium stationary distribution. Due to the causality, x_{n+1} is determined by X_n through the transition probability $P[x_{n+1}|X_n, \lambda_n]$, which depends on the external parameters at time t_n (i.e., λ_n).We note that $P[x_{n+1}|X_n, \lambda_n]$ represents the probability of realizing x_{n+1} at time t_{n+1} under the condition that the trajectory of x until time t_n is given by X_n. If the dynamics is Markovian, $P[x_{n+1}|X_n, \lambda_n]$ can be replaced by $P[x_{n+1}|x_n, \lambda_n]$.

The probability of trajectory X_n is then given by

$$P[X_n|\Lambda_n] = \prod_{k=0}^{n-1} P[x_{k+1}|X_k, \lambda_k] P_0[x_0] \equiv P[X_n], \tag{8.1}$$

where we write $P[X_n|\Lambda_n]$ just as $P[X_n]$ for simplicity. We note that

$$P[X_n|x_0, \Lambda_n] = \prod_{k=0}^{n-1} P[x_{k+1}|X_k, \lambda_k] \equiv P[X_n|x_0] \tag{8.2}$$

is the probability of trajectory X_n under the condition that the initial state is x_0 and the control protocol is Λ_n.

Let A be an arbitrary physical quantity that can depend on the trajectory X_N and protocol Λ_N such that $A[X_N, \Lambda_N]$. The ensemble average of this quantity is given by

$$\langle A \rangle = \int dX_N P[X_N|\Lambda_N] A[X_N, \Lambda_N], \tag{8.3}$$

where $dX_N \equiv \prod_{n=0}^{N} dx_n$.

8.2 Nonequilibrium Equalities

8.2.1 Backward Control

To discuss the nonequilibrium equalities, we first consider the stochastic dynamics with backward control protocols. The backward control protocol means the time-reversal of protocol Λ_N, which is formulated as follows. Let λ^* be the time-reversal of λ; for example, if λ is the magnetic field, then $\lambda^* = -\lambda$. The time-reversed protocol of $\lambda(t)$ is then given by $\lambda^\dagger(t) \equiv \lambda^*(\tau - t)$. The backward protocol can be discretized as $\Lambda_n^\dagger \equiv (\lambda_{N-1}^*, \lambda_{N-2}^*, \ldots, \lambda_{N-n-1}^*)$. We write $\lambda_n^\dagger \equiv \lambda_{N-n-1}^*$, $\lambda_{\mathrm{int}}^\dagger \equiv \lambda_{\mathrm{fin}}^*$, and $\lambda_{\mathrm{fin}}^\dagger \equiv \lambda_{\mathrm{int}}^*$ (see also Fig. 8.1).

We consider the probability of realizing trajectory $x'(t)$ of the system with a backward control protocol. Let $x'_n \equiv x'(n\Delta t)$ and $X'_n = (x'_0, x'_1, \ldots, x'_N)$. We write as $P_0^\dagger[x'_0]$ the initial distribution of the backward processes. We stress that $P_0^\dagger[x'_0]$ is not necessarily equal to the final distribution of the forward experiments. In fact, we can prepare new samples of the system to perform the backward experiments after the forward experiments. The probability distribution of trajectory X'_n with backward protocol is given by

$$P[X'_N|\Lambda_N^\dagger] = \prod_{k=0}^{N-1} P[x'_k|X'_k, \lambda_k^\dagger] P_0^\dagger[x'_0] \equiv P^\dagger[X'_N], \tag{8.4}$$

where we write $P[X'_N|\Lambda_N^\dagger]$ as $P^\dagger[X'_N]$ for simplicity. Correspondingly,

$$P[X'_N|x'_0, \Lambda_N^\dagger] = \prod_{k=0}^{N-1} P[x'_k|X'_k, \lambda_k^\dagger] \equiv P^\dagger[X'_N|x'_0]. \tag{8.5}$$

In special cases, the backward trajectory X'_N is equal to the time-reversal of the forward trajectory X_N. Let x^* be the time-reversal of phase-space point x. For example, if $x = (\boldsymbol{r}, \boldsymbol{p})$ with $\boldsymbol{r}$ and $\boldsymbol{p}$ being the position and the momentum respectively, we have $x^* \equiv (\boldsymbol{r}, -\boldsymbol{p})$. The time-reversal of trajectory X_n is then given by $X_n^\dagger \equiv (x_N^*, x_{N-1}^*, \ldots, x_{N-n}^*)$. With notation $x_n^\dagger \equiv x_{N-n}^*$, we write

$X_n^\dagger = (x_0^\dagger, x_1^\dagger, \ldots, x_n^\dagger)$. By substituting $x_n' = x_n^\dagger$ to Eqs. (8.4) and (8.5), we obtain the probability of realizing a backward trajectory under the backward protocol as

$$P^\dagger[X_N^\dagger] = \prod_{k=0}^{N-1} P[x_k^\dagger | X_k^\dagger, \lambda_k^\dagger] P_0^\dagger[x_0^\dagger], \tag{8.6}$$

and

$$P^\dagger[X_N^\dagger | x_0^\dagger] = \prod_{k=0}^{N-1} P[x_k^\dagger | X_k^\dagger, \lambda_k^\dagger]. \tag{8.7}$$

We note that $dX_N^\dagger = dX_N$ holds, because $dx_n = dx_n^*$.

8.2.2 Formulation of Nonequilibrium Equalities

We now discuss the nonequilibrium equalities. Let $Q_i[X_N, \Lambda_N]$ be the heat that is absorbed by the system from the ith heat bath, which satisfies $Q_i[X_N, \Lambda_N] = -Q_i[X_N^\dagger, \Lambda_N^\dagger]$. We write $Q_i[X_N, \Lambda_N]$ just as $Q_i[X_N]$ for simplicity. It has been established that the following equality is satisfied [4–15, 17–20]:

$$\frac{P^\dagger[X_N^\dagger | x_0^\dagger]}{P[X_N | x_0]} = \exp\left(\sum_i \beta_i Q_i[X_N]\right), \tag{8.8}$$

which is referred to as the detailed fluctuation theorem (or the transient fluctuation theorem). This is the starting point of our research in Chap. 9.

Various proofs of Eq. (8.8) for stochastic systems have been presented, for example, in Refs. [8, 9, 16] for the Markovian stochastic dynamics and in Refs. [17, 18] for non-Markovian dynamics. A proof of Eq. (8.8) has also been given in Ref. [12] for the situations in which the total system including heat baths is treated as a Hamiltonian system. This proof can confirm the physical validity of the detailed fluctuation theorem even for the non-Markovian dynamics with multi-heat baths, as the stochastic dynamics can be reproduced as that of a partial system of the total Hamiltonian system including the heat baths. We also note that several equalities that are similar but not equivalent to Eq. (8.8) have been derived for different situations. For example, the transient fluctuation theorem has been discussed for dynamical systems in Ref. [13]. The fluctuation theorem for nonequilibrium steady states has been discussed for stochastic systems [7, 10] and dynamical systems [4, 5].

We can rewrite Eq. (8.8) as

$$\frac{P^\dagger[X_N^\dagger]}{P[X_N]} = e^{-\sigma[X_N]}, \tag{8.9}$$

where

$$\sigma[X_N] \equiv -\ln P_0^\dagger[x_0^\dagger] + \ln P_0[x_0] - \sum_i \beta_i Q_i[X_N], \tag{8.10}$$

which is called the entropy production along trajectory X_N.

From Eq. (8.9), we can show Crooks' fluctuation theorem [8, 9] as follows. We write as $P[\sigma]$ the probability of obtaining σ in the forward processes, satisfying

$$P[\sigma] = \int \delta(\sigma - \sigma[X_N]) P[X_N] dX_N, \tag{8.11}$$

where $\delta(\cdot)$ is the delta function. On the other hand, let $P^\dagger[\sigma]$ be the probability of obtaining σ in the backward processes, satisfying

$$P^\dagger[\sigma] = \int \delta(\sigma - \sigma[X'_N]) P^\dagger[X'_N] dX'_N. \tag{8.12}$$

By using the detailed fluctuation theorem (8.9) and equality $\sigma[X_N] = -\sigma[X_N^\dagger]$, we obtain

$$\begin{aligned} P[\sigma] &= \int \delta(\sigma - \sigma[X_N]) P[X_N] dX_N \\ &= \int \delta(\sigma + \sigma[X_N^\dagger]) P^\dagger[X_N^\dagger] e^{\sigma[X_N]} dX_N^\dagger \\ &= e^\sigma \int \delta(\sigma + \sigma[X_N^\dagger]) P^\dagger[X_N^\dagger] dX_N^\dagger \\ &= e^\sigma P^\dagger[-\sigma]. \end{aligned} \tag{8.13}$$

Therefore we obtain Crooks' fluctuation theorem

$$\frac{P^\dagger[-\sigma]}{P[\sigma]} = e^{-\sigma}, \tag{8.14}$$

which implies that observations of negative entropy productions in the backward processes are an exponentially rare event compared with the observations of positive entropy productions in the forward process. In special cases in which the system is in a nonequilibrium steady state in which the value of parameter λ is fixed, the backward protocol is the same as the forward one. In such cases, Eq. (8.14) reduces to

$$\frac{P_{\rm ss}[-\sigma]}{P_{\rm ss}[\sigma]} = e^{-\sigma}, \tag{8.15}$$

where $P_{\rm ss}[\sigma]$ is the probability density of σ under the nonequilibrium steady state.

The detailed fluctuation theorem (8.9) or Crooks' fluctuation theorem (8.14) leads to the integral fluctuation theorem

$$\langle e^{-\sigma} \rangle = 1, \tag{8.16}$$

where the ensemble average $\langle \cdots \rangle$ is taken over all trajectories under forward protocol by Eq. (8.3). From the concavity of the exponential function, we obtain

$$\langle \sigma \rangle \geq 0, \tag{8.17}$$

which is an expression of the second law of thermodynamics: the entropy production is non-negative. By taking the ensemble average of the logarithm of both sides of Eq. (8.9), we have a more detailed expression of the second law (8.17):

$$\langle \sigma \rangle = \int dX_N P[X_N] \ln \frac{P[X_N]}{P^\dagger[X_N^\dagger]}, \tag{8.18}$$

which we will refer to as the kawai-parrondo-broeck (KPB) equality [19]. The right-hand side of Eq. (8.18) is the Kullback-Leibler divergence (or the relative entropy) of $P[X_N]$ and $P^\dagger[X_N^\dagger]$, which is always positive. Therefore Eq. (8.18) reproduces inequality (8.17).

If the probability distribution of σ is Gaussian, the cumulant expansion of Eq. (8.3) leads to a variant of fluctuation-dissipation relation

$$\langle \sigma \rangle = \frac{1}{2}(\langle \sigma^2 \rangle - \langle \sigma \rangle^2), \tag{8.19}$$

which means that dissipation $\langle \sigma \rangle$ is determined by the fluctuation of σ. Equality (8.19) is an expression of the fluctuation-dissipation theorem of the first kind, which gives the Green-Kubo formula for special cases [13].

We next discuss the physical meanings of the entropy production in the following three setups.

Isothermal processes. We assume that there is a single heat bath at temperature $T = (k_B\beta)^{-1}$, and that the initial distributions of both the forward and backward experiments are in the canonical distributions. We stress that we do not assume that the final distributions of both the forward and backward experiments are in the canonical distributions: the final distribution of the forward (backward) experiments does not necessarily equal the initial distribution of the backward (forward) experiments. Let $H(x, \lambda)$ be the Hamiltonian of the system with the time symmetry $H(x, \lambda) = H(x^*, \lambda^*)$. The canonical distribution with parameter λ is given by

$$P_{\mathrm{can}}[x|\lambda] \equiv e^{\beta(F(\lambda) - H(x,\lambda))}, \tag{8.20}$$

where

$$F(\lambda) \equiv -k_B T \ln \int dx e^{-\beta H(x,\lambda)} \tag{8.21}$$

is the Helmholtz free energy. In this situation, the entropy production reduces to

$$\sigma[X_N] = \beta(W[X_N] - (\Delta F), \tag{8.22}$$

where

$$W[X_N] \equiv H(x_N, \lambda_{\rm fin}) - H(x_0, \lambda_{\rm int}) - Q[X_N] \tag{8.23}$$

is the work performed on the system from the external parameter, and $\Delta F \equiv F(\lambda_{\rm fin}) - F(\lambda_{\rm int})$ is the free-energy difference.

In this case, Eq. (8.16) leads to

$$\langle e^{-\beta W} \rangle = e^{-\beta \Delta F}, \tag{8.24}$$

which is the Jarzynski equality [6]. With identification $\sigma = \beta(W - \Delta F)$, (8.17) and (8.19) reduces to

$$\langle W \rangle \geq \Delta F \tag{8.25}$$

and

$$\langle W \rangle - \Delta F = \frac{1}{2}(\langle W^2 \rangle - \langle W \rangle^2), \tag{8.26}$$

respectively. Equality (8.26) means that the dissipation of the work ($\langle W \rangle - \Delta F$) is determined by the fluctuation of W.

Multi-heat baths with the initial thermal equilibrium. We assume that there are several heat baths, and that we can control the strength of interaction between the system and the baths through λ. In other words, we can attach or detach the system from the baths by controlling λ; for example, we can attach an adiabatic wall to the system arbitrarily. We assume that, in the initial states of the forward and backward experiments, the system is in contact with a single heat bath at temperature $T = (k_{\rm B}\beta)^{-1}$ and detached from the other baths, and the system is in the canonical distribution. Then the entropy production is given by

$$\begin{aligned} \sigma[X_N] = \beta(H(x_N, \lambda_{\rm fin}) - H(x_0, \lambda_{\rm int})) \\ -\beta\Delta F \quad \sum_i \beta_i Q_i[X_N]. \end{aligned} \tag{8.27}$$

In this case, the second law (8.17) reduces to inequality (5.43).

Transition between arbitrary nonequilibrium states. We set an arbitrary initial distribution $P_0[x_0]$ for the forward experiments. On the other hand, the initial state of the backward experiments is assumed to be taken as $P_0^\dagger[x_0^\dagger] \equiv P_N[x_N]$, where $P_N[x_N]$ is the final distribution of the forward experiments. Although this choice of the backward initial state is artificial and is difficult to be experimentally realized except for special cases, this backward initial state is a theoretically useful tool to derive a version of the second law of thermodynamics as follows. In this case, the entropy production is given by

$$\sigma[X_N] = -\ln P_N[x_N] + \ln P_0[x_0] - \sum_i \beta_i Q_i[X_N], \tag{8.28}$$

and its ensemble average leads to

$$\langle\sigma\rangle = S_N - S_0 - \sum_i \beta_i \langle Q_i\rangle, \tag{8.29}$$

where

$$S_n \equiv -\int P_n[x_n] \ln P_n[x_n] dx_n \tag{8.30}$$

is the Shannon entropy at time t_n. By introducing notation $\Delta S \equiv S_N - S_0$, the second law (8.17) leads to

$$\Delta S \geq \sum_i \beta_i \langle Q_i\rangle, \tag{8.31}$$

which is inequality (5.41). In a special case of a nonequilibrium steady state with two heat baths, inequality (8.31) leads to (5.43).

8.2.3 General Derivation of the Detailed Fluctuation Theorem

We now prove the detailed fluctuation theorem (8.9) for the most general setup. We consider a thermodynamic process of classical system **S** that can contact heat baths $\mathbf{B}_1$, $\mathbf{B}_2$, ..., $\mathbf{B}_n$, at respective temperatures $T_1 = (k_B\beta_1)$, $T_2 = (k_B\beta_2)$, ..., $T_n = (k_B\beta_n)$. To derive the detailed fluctuation theorem (8.9), we treat the total of **S** and $\mathbf{B}_m$'s is a Hamiltonian system. The stochastic feature of the dynamics of **S** is induced by the initial distribution of the heat baths.

Let x be the phase-space point of **S**, and z_m be that of $\mathbf{B}_m$. The total Hamiltonian can be written as

$$H(x, \{z_m\}; \lambda, \{c_m\}) = H^{\mathrm{S}}(x, \lambda) + \sum_{m=1}^{n} (H^{\mathrm{SB}_m}(x, z_m; c_m) + H^{\mathrm{B}_m}(z_m)), \tag{8.32}$$

where $H^{\mathrm{S}}(x; \lambda)$ is the Hamiltonian of **S**, $H^{\mathrm{SB}_m}(x, z_m; c_m)$ is the interaction Hamiltonian between **S** and $\mathbf{B}_m$, and $H^{\mathrm{B}_m}(z_m)$ is the Hamiltonian of $\mathbf{B}_m$. We note that λ describes controllable external parameters, and c_m describes external parameters to control the interaction between **S** and $\mathbf{B}_m$. We assume that the Hamiltonian satisfies the time-reversal symmetry:

$$H(x, \{z_m\}; \lambda, \{c_m\}) = H(x^*, \{z_m^*\}; \lambda^*, \{c_m^*\}). \tag{8.33}$$

We consider a time evolution from 0 to τ, and assume that $H^{\mathrm{SB}_m}(x, z_m; c_m(0)) = H^{\mathrm{SB}_m}(x, z_m; c_m(\tau)) = 0$ holds for all m. We write $H^{\mathrm{S}}(x; \lambda(0)) \equiv H_{\mathrm{i}}^{\mathrm{S}}(x)$, $H^{\mathrm{S}}(x; \lambda(\tau)) \equiv H_{\mathrm{f}}^{\mathrm{S}}(x)$, $H(x, \{z_m\}; \lambda(0), \{c_m(0)\}) \equiv H_{\mathrm{i}}(x, \{z_m\})$, and $H(x, \{z_m\}; \lambda(\tau), \{c_m(0)\}) \equiv H_{\mathrm{f}}(x, \{z_m\})$.

We first consider the forward process with forward protocol $(\lambda(t), \{c_m(t)\})$. Let $X_\tau \equiv \{x(t)\}_{t\in[0,\tau]}$ be a forward trajectory of **S** with initial state $x_0 \equiv x(0)$. We write $z(t) \equiv (z_1(t), z_2(t), \ldots, z_n(t))$ and $z(0) \equiv z_0$. Let Δx_0 and $\Delta z_0 \equiv \Delta z_1(t)\Delta z_2(t)\cdots\Delta z_n(0)$ be the infinitesimal elements in the phase space. We assume that the initial probability of realizing (x_0, z_0) in the forward process is given by

$$P_0[x_0, z_0]\Delta x_0 \Delta z_0 \equiv P_0^{\mathrm{S}}[x_0] P_{\mathrm{can}}^{\mathrm{B}_1}[z_1(0)] \cdots P_{\mathrm{can}}^{\mathrm{B}_n}[z_n(0)]\Delta x_0 \Delta z_0, \tag{8.34}$$

where $P_0^{\mathrm{S}}[x]$ is an arbitrary initial distribution of **S** in the forward process, and

$$P_{\mathrm{can}}^{\mathrm{B}_m}[z_m] \equiv \frac{e^{-\beta_m H^{\mathrm{B}_m}[z_m]}}{Z^{\mathrm{B}_m}} \tag{8.35}$$

is the canonical distribution with $Z^{\mathrm{B}_m} \equiv \int dz_m e^{-\beta H^{\mathrm{B}_m}[z_m]}$. We write the free energies of the heat baths as $F^{\mathrm{B}_m} \equiv -k_{\mathrm{B}} T_m \ln Z^{\mathrm{B}_m}$.

Since the total system is deterministic, X_τ is determined only by the initial state of the total system (x_0, z_0). Therefore the probability of realizing X_τ is given by

$$P[X_\tau]\Delta X_\tau = \sum_{z_0 \text{ (realizes } X_\tau)} P_0[x_0, z_0]\Delta x_0 \Delta z_0, \tag{8.36}$$

where the summation in the right-hand side is over z_0's with which (x_0, z_0) realizes X_τ, and ΔX_τ is the infinitesimal element of X_τ.

We next consider the backward process with the time-reversed control protocol that is given by $(\lambda^\dagger(t), \{c_m^\dagger(t)\}) \equiv (\lambda(\tau - t)^*, \{c_m(\tau - t)^*\})$. Let $X_\tau^\dagger \equiv \{x(\tau - t)^*\}_{t\in[0,\tau]}$ be the time-reversal of X_τ. We write $z_0^\dagger \equiv z(\tau)^*$ We note that the initial state of $X_\tau^\dagger$ is given by $x_0^\dagger = x(\tau)^*$. We assume that the initial distribution of the total system in the backward process is given by

$$P_0^\dagger[x_0^\dagger, z_0^\dagger]\Delta x_0^\dagger \Delta z_0^\dagger \equiv P_0^{\dagger\mathrm{S}}[x_0^\dagger] P_{\mathrm{can}}^{\mathrm{B}_1}[z_1^\dagger(0)] \cdots P_{\mathrm{can}}^{\mathrm{B}_n}[z_n^\dagger(0)]\Delta x_0^\dagger \Delta z_0^\dagger, \tag{8.37}$$

where $P_0^{\dagger\mathrm{S}}[x_0^\dagger]$ is an arbitrary initial distribution of **S** in the backward process.

Since the total system is reversible, the initial state of the backward process that realizes $X_\tau^\dagger$ is given by $(x_0^\dagger, z_0^\dagger) \equiv (x(\tau)^*, z(\tau)^*)$. The corresponding infinitesimal element is given by $\Delta x_0^\dagger \Delta z_0^\dagger \equiv \Delta x(\tau)^* \Delta z(\tau)^*$. From Liouville's theorem, we have

$$\Delta x_0 \Delta z_0 = \Delta x(\tau)\Delta z(\tau). \tag{8.38}$$

Therefore, noting that $\Delta x(\tau)\Delta z(\tau) = \Delta x(\tau)^* \Delta z(\tau)^*$ holds, we obtain

$$\Delta x_0^\dagger \Delta z_0^\dagger = \Delta x_0 \Delta z_0. \tag{8.39}$$

The conditional probability of realizing $X_\tau^\dagger$ with the time-reversed protocol is given by

$$P^\dagger[X_\tau^\dagger]\Delta X_\tau^\dagger = \sum_{z_0^\dagger \text{ (realizes } X_\tau^\dagger)} P_0^\dagger[x_0^\dagger, z_0^\dagger]\Delta x_0^\dagger \Delta z_0^\dagger, \tag{8.40}$$

where the summation in the right-hand side is over $z_0^\dagger$'s with which $(x_0^\dagger, z_0^\dagger)$ realizes $X_\tau^\dagger$, and $\Delta X_\tau^\dagger$ is the infinitesimal element of $X_\tau^\dagger$. We note that $\Delta X_\tau = \Delta X_\tau^\dagger$ is assumed to hold by definition.

On the other hand, by using the time-reversal symmetry $P_{\text{can}}^{\mathrm{B}_m}[z_m] \equiv P_{\text{can}}^{\mathrm{B}_m}[z_m^*]$, we obtain

$$\frac{P_{\text{can}}^{\mathrm{B}_1}[z_1^\dagger(0)] \cdots P_{\text{can}}^{\mathrm{B}_n}[z_n^\dagger(0)]}{P_{\text{can}}^{\mathrm{B}_1}[z_1(0)] \cdots P_{\text{can}}^{\mathrm{B}_n}[z_n(0)]} = \exp\left(\sum_m \beta_m Q_m[x_0, z_0]\right), \tag{8.41}$$

where

$$Q_m[x_0, z_0] \equiv H^{\mathrm{B}_m}[z_m(0)] - H^{\mathrm{B}_m}[z_m(\tau)] \tag{8.42}$$

is the energy-difference of $\mathbf{B}_m$, that is, the heat that is absorbed in $\mathbf{S}$ from $\mathbf{B}_m$. We note that the ratio of the probability densities is related to the heats due to the initial canonical distributions of the heat baths.

From Eqs. (8.36), (8.39–8.41), we obtain

$$\begin{aligned} &P^\dagger[X_\tau^\dagger]\Delta X_\tau^\dagger \\ &= \sum_{z_0 \text{ (realizes } X_\tau)} P_0^{\dagger\mathrm{S}}[x_0^\dagger] P_{\text{can}}^{\mathrm{B}_1}[z_1(0)] \cdots P_{\text{can}}^{\mathrm{B}_n}[z_n(0)] \exp\left(\sum_m \beta_m Q_m[x_0, z_0]\right) \Delta x_0 \Delta z_0 \\ &= \sum_{z_0 \text{ (realizes } X_\tau)} P_0[x_0, z_0]\Delta x_0 \Delta z_0 \exp\left(\ln P_0^{\dagger\mathrm{S}}[x_0^\dagger] - \ln P_0^{\mathrm{S}}[x_0] + \sum_m \beta_m Q_m[x_0, z_0]\right). \end{aligned} \tag{8.43}$$

We then make a crucial assumption that $Q_m[x_0, z_0]$ is determined only by the trajectory of $\mathbf{S}$, i.e., $Q_m[x_0, z_0] = Q_m[X_\tau]$. This assumption is valid for many physical systems; for example, Langevin systems satisfy this condition. We note that x_0 and $x_0^\dagger \equiv x(\tau)^*$ are determined only by X_τ. By introducing the entropy production as

$$\sigma[X_\tau] \equiv -\ln P_0^{\dagger\mathrm{S}}[x_0^\dagger] + \ln P_0^{\mathrm{S}}[x_0] - \sum_m \beta_m Q_m[X_\tau], \tag{8.44}$$

we obtain

$$P^\dagger[X_\tau^\dagger]\Delta X_\tau^\dagger = e^{-\sigma[X_\tau]} \sum_{z_0 \text{ (realizes } X_\tau)} P_0[x_0, z_0]\Delta x_0 \Delta z_0 = e^{-\sigma[X_\tau]} P[X_\tau]\Delta X_\tau, \tag{8.45}$$

which leads to

$$\frac{P^\dagger[X_\tau^\dagger]}{P[X_\tau]} = e^{-\sigma[X_\tau]}. \tag{8.46}$$

Equality (8.46) is the detailed fluctuation theorem (8.9).

8.3 Markovian Dynamics with Detail Balance

As an example, we consider the Markovian dynamics with a single heat bath at $T = (k_B\beta)^{-1}$. We assume that there is not any non-conservative force, so that the dynamics satisfies the detail balance. In this subsection, we assume that $\lambda = \lambda^*$ holds.

8.3.1 General Formulation

In general, the master equation in the continuous time can be written as

$$\frac{dP[x,t]}{dt} = -\int p[x'|x, \lambda(t)]P[x,t]dx' + \int p[x|x', \lambda(t)]P[x,t]dx', \tag{8.47}$$

where $P[x,t]$ is the probability density at time t_n and $p[x'|x,\lambda]$ is the transition rate per unit time from x to x' under parameter λ. Equation (8.47) can be discretized as

$$P_{n+1}[x] = P_n[x] - \Delta t \int p[x'|x, \lambda_n]P_n[x]dx' + \Delta t \int p[x|x', \lambda_n]P_n[x]dx', \tag{8.48}$$

where $P_n[x] \equiv P[x, n\Delta t]$ and $\Lambda_n \equiv \lambda(n\Delta t)$. We then obtain the transition probability from x_n to x_{n+1} as

$$P[x_{n+1}|x_n, \lambda_n] = \delta(x_{n+1} - x_n)\left(1 - \Delta t \int p[x'|x_n, \lambda_n]dx'\right) + \Delta t p[x_{n+1}|x_n, \lambda_n], \tag{8.49}$$

where $\delta(\cdots)$ is the delta function. Then the master equation (8.47) is equivalent to

$$P_{n+1}[x_{n+1}] = \int dx_n P[x_{n+1}|x_n, \lambda_n]P[x_n]. \tag{8.50}$$

In general, let $P[x'|x,\lambda]$ the transition probability from x to x' under parameter λ. The detailed balance means that, in the thermal equilibrium, there is no net flow between any two phase-space points, which is described as, for arbitrary x, x', λ,

$$P[x'|x,\lambda]\exp\left(-\beta H^{\rm S}(x,\lambda)\right)=P[x^*|x'^*,\lambda]\exp\left(-\beta H^{\rm S}(x',\lambda)\right), \tag{8.51}$$

where we used the time-reversal symmetry of the Hamiltonian: $H^{\rm S}(x',\lambda)=H^{\rm S}(x^*,\lambda)$. From Eq. (8.51), we obtain

$$\frac{P[x_n^*|x_{n+1}^*,\lambda_n]}{P[x_{n+1}|x_n,\lambda_n]}=e^{\beta Q[x_n,x_{n+1}]}, \tag{8.52}$$

where

$$Q[x_n,x_{n+1}]\equiv H(x_{n+1},\lambda_n)-H(x_n,\lambda_n) \tag{8.53}$$

is the heat that is absorbed in the system between time interval n to $n+1$ in which x_n transits to x_{n+1}.

Since the dynamics is Markovian, the probability density of trajectory X_N under the condition of x_0 with the forward protocol is given by

$$P[X_N|x_0]=\prod_{n=0}^{N-1}P[x_{n+1}|x_n,\lambda_n], \tag{8.54}$$

while with the backward protocol,

$$P[X_N^\dagger|x_0^\dagger]=\prod_{n=0}^{N-1}P[x_{n+1}^\dagger|x_n^\dagger,\lambda_n^\dagger]. \tag{8.55}$$

We note that

$$P[x_n^*|x_{n+1}^*,\lambda_n]=P[x_{N-n}^\dagger|x_{N-n-1}^\dagger,\lambda_{N-n-1}^\dagger] \tag{8.56}$$

Therefore, by taking the product from (8.52) in terms of n, we obtain the detailed fluctuation theorem (8.8):

$$\frac{P^\dagger[X_N^\dagger|x_0^\dagger]}{P[X_N|x_0]}=e^{\beta Q[X_N]}, \tag{8.57}$$

where $Q[X_N]=\sum_{n=0^{N-1}}Q[x_n,x_{n+1}]$. Therefore, in the Markovian dynamics, the detailed fluctuation theorem reduces to a straightforward consequence of the detailed balance. However, we stress that the original detailed fluctuation theorem (8.8) is a far more general statement, which is valid even without the detailed balance.

Corresponding to the definition of the heat (8.53), we define the work as

$$W[x_n, x_{n+1}] \equiv H(x_n, \lambda_{n+1}) - H(x_n, \lambda_n), \tag{8.58}$$

which leads to the first law of thermodynamics

$$H(x_{n+1}, \lambda_{n+1}) - H(x_n, \lambda_n) = W[x_n, x_{n+1}] + Q[x_n, x_{n+1}] + o(\Delta t). \tag{8.59}$$

8.3.2 Overdamped Langevin Systems

As the simplest example of the Markovian dynamics with the detailed balance, we discuss a Brownian particle with position x. Let $V(x, \lambda)$ be the potential and $F(x, \lambda)$ be the force satisfying

$$F(x, \lambda) = -\frac{\partial V(x, \lambda)}{\partial x}. \tag{8.60}$$

If the time scale Δt is much larger than the inverse of the friction constant, the system obeys the overdamped Langevin equation, in which we can neglect the term of the momentum, because it relaxes much faster than the position. In typical actual experiments with, for example, a colloidal particle, the overdamped Langevin equation is valid.

The overdamped Langevin equation is given

$$\eta \frac{dx(t)}{dt} = F(x(t), \lambda(t)) + \sqrt{2\eta k_B T}\xi(t), \tag{8.61}$$

where η is the friction constant, $\xi(t)$ is the Gaussian white noise satisfying $\langle \xi(t)\xi(t')\rangle = \delta(t - t')$ with $\delta(\cdot)$ being the delta function. We then consider the discretization of Eq. (8.61). In this case, the Itô and the Stratonovich interpretations are mathematically equivalent, because $\sqrt{2\eta k_B T}$ is independent of x. The Itô interpretation is given by

$$\eta(x_{n+1} - x_n) = F_n \Delta t + \sqrt{2\eta k_B T}\Delta W_n, \tag{8.62}$$

while the Stratonovich one by

$$\eta(x_{n+1} - x_n) = \frac{F_{n+1} + F_n}{2}\Delta t + \sqrt{2\eta k_B T}\Delta W_n, \tag{8.63}$$

where F_n is given by $F(x_n, \lambda_n)$ and ΔW_n is the Wiener increment satisfying $\Delta W_n \Delta W_{n'} = \delta_{nn'}\Delta t$. The transition probability from x_n to x_{n+1} is given by

$$P[x_{n+1}|x_n, \lambda_n] \propto \exp\left[-\frac{\eta\beta}{4\Delta t}\left(x_{n+1} - x_n - \frac{F_n}{\eta}\Delta t\right)^2\right], \tag{8.64}$$

which is also independent of the interpretations. The corresponding master equation is given by the Fokker-Planck equation:

$$\frac{\partial P[x,t]}{\partial t} = \frac{1}{\eta}\frac{\partial}{\partial x}\left(\frac{\partial V}{\partial x} + k_{\mathrm{B}}T\frac{\partial}{\partial x}\right)P[x,t]. \tag{8.65}$$

We note that Eq. (8.64) can be obtained as the discretization of Eq. (8.65).

The heat is given by

$$\begin{aligned} Q[x_n, x_{n+1}] &\equiv V(x_{n+1}, \lambda_n) - V(x_n, \lambda_n) \\ &= \left(-\eta\frac{x_{n+1} - x_n}{\Delta t} + \sqrt{2\eta k_{\mathrm{B}}T}\frac{\Delta W_n}{\Delta t}\right)(x_{n+1} - x_n) + o(\Delta t) \\ &= -\frac{F_n + F_{n+1}}{2}(x_{n+1} - x_n) + o(\Delta t), \end{aligned} \tag{8.66}$$

which should be in the Stratonovich form [1]. The work is given by

$$W_n \equiv V(x_n, \lambda_{n+1}) - V(x_n, \lambda_n). \tag{8.67}$$

The detailed balance is satisfied for the Langevin system:

$$\frac{P[x_n|x_{n+1}, \lambda_n]}{P[x_{n+1}|x_n, \lambda_n]} = e^{\beta Q[x_n, x_{n+1}]}. \tag{8.68}$$

Therefore, we obtain the detailed fluctuation theorem

$$\frac{P^\dagger[X_N^\dagger|x_0^\dagger]}{P[X_n|x_0]} = e^{\beta Q[X_N]}, \tag{8.69}$$

where $Q[X_N] \equiv \sum_{n=0}^{N-1} Q[x_n, x_{n+1}]$.

References

1. K. Sekimoto, Prog. Theor. Phys. Suppl. **130**, 17 (1998)
2. C. Bustamante, J. Liphardt, F. Ritort, Phys. Today **58**, 43 (2005)
3. U. Seifert, Euro. Phys. J. B **64**, 423 (2008)
4. D.J. Evans, E.G.D. Cohen, G.P. Morriss, Phys. Rev. Lett. **71**, 2401 (1993)
5. G. Gallavotti, E.G.D. Cohen, Phys. Rev. Lett. **74**, 2694 (1995)
6. C. Jarzynski, Phys. Rev. Lett. **78**, 2690 (1997)
7. J. Kurchan, J. Phys. A: Math. Gen. **31**, 3719 (1998)
8. G.E. Crooks, J. Stat. Phys. **90**, 1481 (1998)

9. G.E. Crooks, Phys. Rev. E **60**, 2721 (1999)
10. J.L. Lebowitz, H. Spohn, J. Stat. Phys. **95**, 333 (1999)
11. C. Maes, J. Stat. Phys. **95**, 367 (1999)
12. C. Jarzynski, J. Stat. Phys. **98**, 77 (2000)
13. D.J. Evans, D.J. Searles, Adv. Phys. **51**, 1529 (2002)
14. R. van Zon, E.G.D. Cohen, Phys. Rev. Lett. **91**, 110601 (2003)
15. D. Andrieux, P. Gaspard, J. Chem. Phys. **121**, 6167 (2004)
16. U. Seifert, Phys. Rev. Lett. **95**, 040602 (2005)
17. T. Ohkuma, T. Ohta, J. Stat. Mech. **P10010**, (2007)
18. T. Speck, U. Seifert, J. Stat. Mech. **L09002**, (2007)
19. R. Kawai, J.M.R. Parrondo, C. Van den Broeck, Phys. Rev. Lett. **98**, 080602 (2007)
20. S. Vaikuntanathan, C. Jarzynski, Euro. Phys. Lett. **87**, 60005 (2010)
21. J. Liphardt et al., Science **296**, 1832 (2002)
22. D. Collin et al., Nature **437**, 231 (2005)
23. T. Sagawa, M. Ueda, Phys. Rev. E **85**, 021104 (2012)

Chapter 9
Nonequilibrium Equalities with Feedback Control

Abstract In this chapter, we consider the stochastic aspects of thermodynamics of feedback control, by taking into account the fluctuations of thermodynamic quantities and information contents. In particular, we generalize the nonequilibrium equalities such as the fluctuation theorem and the Jarzynski equalities to feedback-controlled classical stochastic systems [1, 2], which is the third main part of this thesis. This topic is related to the paradox of Maxwell's demon [3–7], and has been a topic of active researches in terms of modern nonequilibrium statistical mechanics [1, 2, 8–31]. In Sect. 9.1, we formulate measurements on nonequilibrium systems. In Sect. 9.2, we formulate the effect of feedback control. In Sect. 9.3, we derive the main results in this chapter, which are the generalized nonequilibrium equalities. In Sect. 9.4, we discuss two typical examples that illustrate our general results.

9.1 Effect of Measurements

First of all, we formulate multiple measurements on nonequilibrium dynamics, and discuss the properties the mutual information obtained by the measurements.

9.1.1 Formulation

Let y_n be the outcome at time $t_n \equiv n\Delta t$. In this section, we assume the followings:

1. The error of the measurement at time t_n is characterized by $P[y_n|X_n]$, where y_n can depend on the trajectory of the system before n due to the causality. Here we assumed that the property of the measurement error at time t_n does not explicitly depend on Y_{n-1} nor $P[X_n]$. This assumption is also justified in many real experimental situations.

T. Sagawa, *Thermodynamics of Information Processing in Small Systems*,
Springer Theses, DOI: 10.1007/978-4-431-54168-4_9, © Springer Japan 2013

2. We can neglect the back-action of the measurement: the unconditional probability distribution of X_n, $P[X_n]$, is not affected irrespective of whether we perform measurements or not. Since the system is classical, this assumption is justified for many real systems such as colloidal particles and macromolecules.

If $P[y_n|X_n] = P[y_n|x_n]$ holds, the measurement is called Markovian, which means the measurements can be performed in a time interval that is sufficiently shorter than the shortest time scale Δt of the system. We note that the Markovness of the measurement is independent of that of the dynamics.

We assume that the measurements are performed at times $t_{n_1}, t_{n_2}, \ldots, t_{n_M}$, where $0 \leq n_1 < n_2 < \cdots < n_M \leq N$. If $n_1 = 0, n_2 = 1, n_3 = 2, \ldots, n_{N+1} = N$ hold, the measurement is time-continuous.

We write as Y_n the set of measurement outcomes that are obtained up to time t_n, i.e. $Y_n \equiv (y_{n_1}, y_{n_2}, \ldots, y_{[n]})$ where $[n]$ is the maximum n_k satisfying $n_k \leq n$. If the measurement is continuous, then $Y_n = (y_0, y_1, \ldots, y_n)$.

We define

$$P_{\mathrm{c}}[Y_n|X_n] \equiv \prod_{k=1}^{M_n} P[y_{n_k}|X_{n_k}], \tag{9.1}$$

where M_n is the maximum integer satisfying $n_{M_n} \leq n$. Without feedback, Eq. (9.1) defines the conditional probability of obtaining outcomes Y_n under the condition of X_n, while, with feedback, this interpretation of Eq. (9.1) is not necessarily correct as shown in the next section. To explicitly demonstrate this point and to distinguish $P_{\mathrm{c}}[Y_n|X_n]$ from the usual conditional probability, we put suffix "c". Then the joint distribution of X_n and Y_n is given by

$$P[X_n, Y_n] = P_{\mathrm{c}}[Y_n|X_n]P[X_n], \tag{9.2}$$

where we used the aforementioned assumption that we can neglect the back-action of measurements. The probability of obtaining outcomes Y_n is given by

$$P[Y_n] = \int dX_n P[X_n, Y_n] = \prod_{k=1}^{M_n} P[y_{n_k}|Y_{n_k - 1}], \tag{9.3}$$

where the two equalities are just identities known in probability theory. We also note that

$$\begin{aligned} P[y_n|X_n, Y_{n-1}] &\equiv \frac{P[Y_n|X_n]}{P[Y_{n-1}|X_n]} \\ &= P[y_n|X_n], \end{aligned} \tag{9.4}$$

which is in fact independent of Y_{n-1}.

9.1.2 Mutual Information

We next consider the mutual information that is obtained by measurements. In general, let x be a measured state and y be a measurement outcome. In this chapter, we use the notation

$$I[x:y] \equiv \ln \frac{P[y|x]}{P[y]}, \tag{9.5}$$

whose ensemble average leads to the mutual information

$$\langle I \rangle \equiv \int I[x:y]P[x,y]dxdy = \int \ln \frac{P[x,y]}{P[x]P[y]} P[x,y]dxdy. \tag{9.6}$$

We also refer to $I[x:y]$ as the mutual information.

We then discuss the mutual information obtained by multiple measurements on nonequilibrium dynamics. Suppose that we obtain measurement outcomes Y_{n-1} at time t_{n-1}. If we perform another measurement at time t_n and obtain outcome y_n, we obtain the mutual information between y_n and X_n under the condition that we have had Y_{n-1}:

$$\begin{aligned} I[X_n : y_n|Y_{n-1}] &\equiv \ln \frac{P[y_n|X_n, Y_{n-1}]}{P[y_n|Y_{n-1}]} \\ &= \ln \frac{P[y_n|X_n]}{P[y_n|Y_{n-1}]}, \end{aligned} \tag{9.7}$$

where we used Eq. (9.4). We note that, if the measurement is Markovian, $I[X_n : y_n|Y_{n-1}]$ reduces to $I[x_n : y_n|Y_{n-1}]$. We write as I_c the sum of these mutual information contents obtained by multi-measurements, which is given by

$$\begin{aligned} I_c[X_n : Y_n] &\equiv \sum_{k=1}^{M_n} I[X_{n_k} : y_{n_k}|Y_{n_k-1}] \\ &= \ln \frac{P_c[Y_n|X_n]}{P[Y_n]}, \end{aligned} \tag{9.8}$$

where we used Eq. (9.3). We note that the same quantity has been discussed in Refs. [15, 32]. From Eq. (9.8), we find that $I_c[Y_n : X_n]$ equals the mutual information between trajectories X_n and Y_n defined as $I[Y_n : X_n] \equiv \ln(P[Y_n|X_n]/P[Y_n])$. In the presence of feedback control, however, this is not true (i.e., $I_c \neq I$), as we will see later.

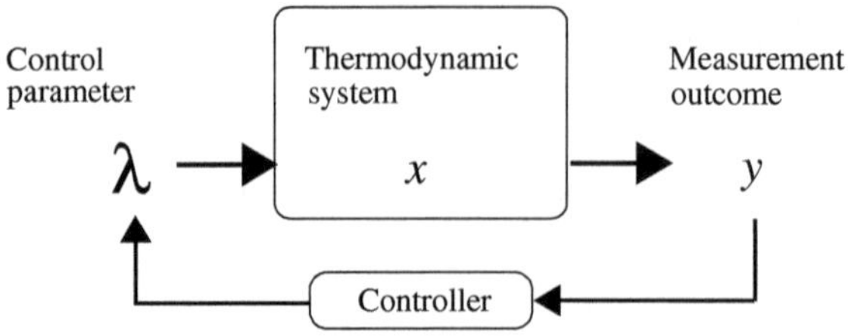

Fig. 9.1 Feedback control on nonequilibrium dynamics (reproduced from Ref. [2] with permission). The control parameter is denoted as λ, the point of the phase space of the system as x, and the outcome of measurement on the system as y. Parameter λ depends on y through the real-time feedback control

9.2 Feedback Control

We now formulate feedback control [33, 34] on nonequilibrium dynamics. Feedback control means that protocol Λ_N depends on measurement outcomes Y_N (see Fig. 9.1). On the other hand, without feedback control, control protocols are pre-fixed and independent of measurement outcomes, as is the case for the setup of the original fluctuation theorem and Jarzynski equality.

9.2.1 Formulation

When the system is subject to feedback control, λ_n can depend on measurement outcomes that are obtained until time t_n, while λ_n cannot depend on any measurement outcome that is obtained after time t_n due to causality. We introduce notation $\lambda_n(Y_n)$, which means that the value of λ at time t_n is determined by Y_n. We write $\Lambda_n(Y_{n-1}) \equiv (\lambda_0(Y_0), \lambda_1(Y_1), \ldots, \lambda_{n-1}(Y_{n-1}))$.

If λ_n only depends on y_n as $\lambda_n(y_n)$, the feedback protocol is called Markovian. We note that the Markovness of feedback is independent of that of dynamics or measurements. The Markovian feedback control is realized when the delay time of feedback is sufficiently smaller than the smallest time scale Δt of the dynamics.

As a simple illustrative example, we discuss an overdamped Langevin system, where the Langevin equation is given by Eq. (8.61). We assume that the measurement is time-continuous and Markovian as

$$y_n = x_n \Delta t + \Delta R_n, \tag{9.9}$$

where ΔR_n is a white Gaussian noise with $\Delta R_n \Delta R_{n'} = R\delta_{nn'}\Delta t$ $(R > 0)$. The conditional probability of obtaining an outcome is given by

$$P[y_n|x_n] \propto \exp\left[-\frac{1}{2R\Delta t}(y_n - x_n\Delta t)^2\right]. \tag{9.10}$$

The feedback protocol can be written as $\lambda_n(y_0, y_1, \ldots, y_n)$ in general.

We note that the work (8.67) is given by

$$W_n = \frac{\partial V}{\partial \lambda} \Delta\lambda_n + o(\Delta t), \tag{9.11}$$

where

$$\Delta\lambda_n \equiv \lambda_{n+1}(y_0, y_1, \ldots, y_{n+1}) - \lambda_n(y_0, y_1, \ldots, y_n). \tag{9.12}$$

In particular, if the feedback is Markovian, λ_n is given by $\lambda_n(y_n)$. Then $\Delta\lambda_n = \lambda_{n+1}(y_{n+1}) - \lambda_n(y_n)$ can be written as

$$\Delta\lambda_n = \frac{\partial \lambda}{\partial t} \Delta t + \frac{\partial \lambda}{\partial y} \Delta y_n + \frac{1}{2} \frac{\partial^2 \lambda}{\partial y^2} \Delta y_n^2, \tag{9.13}$$

where $\Delta y_n \equiv y_{n+1} - y_n$. The first term on the right-hand side of Eq. (9.13) arises from a change in λ by the pre-fixed protocol, while the second and third terms are induced by the feedback control.

We next consider the Kalman filter and the optimal control. As a special case of Eq. (8.61), we consider a linear Langevin equation in the Itô form:

$$\eta(x_{n+1} - x_n) = -K x_n \Delta t + \lambda_n \Delta t + \sqrt{2\eta k_B T} \Delta W_n, \tag{9.14}$$

where K is a positive constant and λ_n is a control parameter. We consider the Kalman filter that is a standard method to construct the optimal estimator of x_n, denoted as $\hat{x}_n$, in terms of the mean square error. From measurement outcomes Y_n, $\hat{x}_n$ is obtained as the solution of the following simultaneous differential equations [35]:

$$\hat{x}_{n+1} - \hat{x}_n = -\frac{K\hat{x}_n + \lambda_n}{\eta} \Delta t + \frac{A_n}{R}(y_n - \hat{x}_n \Delta t), \tag{9.15}$$

$$A_{n+1} - A_n = \left(\frac{-2K A_n + 2k_B T}{\eta} - \frac{A_n}{R} \right) \Delta t, \tag{9.16}$$

where A_n is a time-dependent real number and Eq. (9.16) is a discretized version of the Riccati equation. By using the Kalman estimator $\hat{x}_n$, the optimal control protocol [36] is given by

$$\lambda_n = -C_n \hat{x}_n, \tag{9.17}$$

where C_n is a pre-determined constant depending on the target of the optimal control. We note that the optimal control is a non-Markovian control as $\lambda_n = \lambda_n(Y_n)$, because we use all of $Y_n = (y_0, y_1, \ldots, y_n)$ to calculate $\hat{x}_n$. The generalized Jarzynski equality for this situation has been obtained in Ref. [19].

9.2.2 Probability Distributions with Feedback

We discuss the probability distributions with feedback control in general. Under the condition that we fix control protocol $\Lambda_n(Y_{n-1})$ with Y_{n-1} being fixed, the conditional probability of realizing X_n is given by

$$P[X_n|\Lambda_n(Y_{n-1})] = P_0[X_0]\prod_{k=0}^{n-1} P[x_{k+1}|X_k, \lambda_k(Y_k)], \tag{9.18}$$

which corresponds to Eq. (8.1). We note that, in the expression in Eq. (9.18), we do not omit the notation $\Lambda_N(Y_{N-1})$ because its Y_{N-1}-dependence is crucial. We also write

$$P[X_n|x_0, \Lambda_n(Y_{n-1})] \equiv \prod_{k=0}^{n-1} P[x_{k+1}|X_k, \lambda_k(Y_k)]. \tag{9.19}$$

On the other hand, along the trajectory X_n, the conditional probability of obtaining outcome y_{n_k} at time t_{n_k} is written as $P[y_{n_k}|X_{n_k}]$. We then define $P_c[Y_n|X_n]$ by Eq. (9.1).

We then obtain the joint probability distribution of X_n and Y_n with feedback control as

$$\begin{aligned} P[X_n, Y_n] &= P[x_0]\prod_{k=1}^{M_n} P[y_{n_k}|X_{n_k}]\prod_{k=0}^{n-1} P[x_{k+1}|X_k, \lambda_k(Y_k)] \\ &= P_c[Y_n|X_n]P[X_n|\Lambda_n(Y_{n-1})]. \end{aligned} \tag{9.20}$$

We can check that

$$\int P[X_n, Y_n]dX_n dY_n = 1, \tag{9.21}$$

by integrating X_n and Y_n in Eq. (9.20) in the order of $x_n \to \cdots \to x_{n_{M_n}+1} \to y_{n_{M_n}} \to x_{n_{M_n}-1} \to \cdots \to x_{n_1+1} \to y_{n_1} \to x_{n_1-1} \to \cdots \to x_0$, where the causality of measurements and feedback play crucial roles.

The marginal distributions are given by

$$P[X_n] = \int P[X_n, Y_n]dY_n, \quad P[Y_n] = \int P[X_n, Y_n]dX_n, \tag{9.22}$$

and the conditional distributions by

$$P[X_n|Y_n] = \frac{P[X_n, Y_n]}{P[Y_n]}, \quad P[Y_n|X_n] = \frac{P[X_n, Y_n]}{P[X_n]}. \tag{9.23}$$

We stress that, in the presence of feedback control,

$$P[Y_n|X_n] \neq P_c[Y_n|X_n] \tag{9.24}$$

in general, because protocol Λ_N depends on Y_{N-1}. On the other hand, without feedback control, $P[Y_n|X_n] = P_c[Y_n|X_n]$ holds because $P[X_n]$ is just given by $P[X_n|\Lambda_n]$ with Λ_n being independent of Y_n.

The ensemble average of a probability variable $A[X_n, Y_n]$ is given by

$$\langle A \rangle \equiv \int A[X_n, Y_n] P[X_n, Y_n] dX_n dY_n, \tag{9.25}$$

and the conditional average under the condition of Y_n is given by

$$\langle A \rangle_{Y_n} \equiv \int A[X_n, Y_n] P[X_n|Y_n] dX_n. \tag{9.26}$$

Equation (9.4) still holds in the presence of feedback control:

$$\begin{aligned} P[y_n|X_n, Y_{n-1}] &\equiv \frac{P[X_n, Y_n]}{P[X_n, Y_{n-1}]} \\ &= \frac{P_c[Y_n|X_n] P[X_n|\Lambda_n(Y_{n-1})]}{P_c[Y_{n-1}|X_{n-1}] P[X_n|\Lambda_n(Y_{n-1})]} \\ &= P[y_n|X_n]. \end{aligned} \tag{9.27}$$

We note that Eq. (9.3) also holds with feedback control.

We then define the mutual information in the same way as in the case without feedback control:

$$\begin{aligned} I_c[Y_n : X_n] &= \sum_{k=1}^{M_n} I[y_{n_k} : X_{n_k} | Y_{n_k - 1}] \\ &= \ln \frac{P_c[Y_n|X_n]}{P[Y_n]}. \end{aligned} \tag{9.28}$$

In the presence of feedback control, $I_c[Y_n : X_n]$ does not equal the mutual information between trajectories X_n and Y_n defined as $I[Y_n : X_n] \equiv \ln(P[Y_n|X_n]/P[Y_n])$, because $P_c[Y_n|X_n] \neq P[Y_n|X_n]$. Intuitively speaking, I_c only characterizes the correlation between X_n and Y_n due to the measurements, while I involves the correlation due to the feedback control. Note that I_c is a more important quantity than I, because I_c has a clear information-theoretic significance: I_c is the information that we obtain by measurements. We also note that, in the case of a single measurement and feedback, $I_c = I$ always holds.

We note that an identity similar to the integral fluctuation theorem holds for I_c:

$$\langle e^{-I_c} \rangle = 1, \tag{9.29}$$

because

$$\langle e^{-I_c} \rangle = \int dX_N dY_N \frac{P[Y_N]}{P_c[Y_N|X_N]} P[X_N, Y_N]$$
$$= \int dX_N dY_N P[Y_N] P[X_N|\Lambda_N(Y_{N-1})] = 1. \tag{9.30}$$

9.2.3 Detailed Fluctuation Theorem for a Fixed Control Protocol

If we fix control protocol $\Lambda_N(Y_{N-1})$ with Y_{N-1} being fixed, then the detailed fluctuation theorem (8.8) still holds:

$$\frac{P[X_N^\dagger | x_0^\dagger, \Lambda_N(Y_{N-1})^\dagger]}{P[X_N|x_0, \Lambda_N(Y_{N-1})]} = \exp\left(\sum_i \beta_i Q_i[X_N, \Lambda_N(Y_{N-1})] \right), \tag{9.31}$$

where

$$\Lambda_N(Y_{N-1})^\dagger \equiv (\lambda_{N-1}(Y_{N-1})^*, \ldots, \lambda_0(Y_0)^*). \tag{9.32}$$

The left-hand side of Eq. (9.31) corresponds to the following forward and backward experiments. We first perform forward experiments many times with feedback control, and choose the subensemble in which the measurement outcomes are given by Y_{N-1}. Within this subensemble, the ratio of trajectory X_N is given by $P[X_N|x_0, \Lambda_N(Y_{N-1})]$ under the condition of initial x_0. We next perform backward experiments with protocol $\Lambda_N(Y_{N-1})^\dagger$ where Y_{N-1} was chosen in the forward experiments. We stress that we do not perform any feedback in the backward experiments: $\Lambda_N(Y_{N-1})^\dagger$ is just the time-reversal of $\Lambda_N(Y_{N-1})$. We then obtain $P[X_N^\dagger | x_0^*, \Lambda_N(Y_{N-1})^\dagger]$ as the ratio of trajectory $X_N^\dagger$, under the condition of initial $x_0^\dagger$ in the backward experiments. The original detailed fluctuation theorem (8.8) can straightforwardly be applied to this subensemble corresponding to Y_{N-1} because we have a unique control protocol in the subensemble, and therefore we obtain Eq. (9.31).

We take the initial distribution of the backward experiments as $P_0^\dagger[x_0^\dagger|Y_N]$ that in general depends on the measurement outcomes in the forward experiments. The natural choice of $P_0^\dagger[x_0^\dagger|Y_N]$ is a stationary state $P_s[x_0^\dagger|\lambda_{\rm fin}(Y_N)^*]$. We have

$$\frac{P[X_N^\dagger | \Lambda_N(Y_{N-1})^\dagger]}{P[X_N|\Lambda_N(Y_{N-1})]} = \exp\left(-\sigma[X_N, \Lambda_N(Y_{N-1})]\right), \tag{9.33}$$

where

$$\sigma[X_N, \Lambda_N(Y_{N-1})] \equiv -\ln P_0^\dagger[x_0^\dagger|Y_N] + \ln P_0[x_0]$$
$$- \sum_i \beta_i Q_i[X_N, \Lambda_N(Y_{N-1})]. \tag{9.34}$$

If there is a single heat bath and the initial distributions of the forward and backward experiments are given by the canonical distributions, then the entropy production reduces to

$$\sigma[X_N, \Lambda_N(Y_{N-1})] = \beta(W[X_N, \Lambda_N(Y_{N-1})] - \Delta F[Y_N]), \tag{9.35}$$

where the free-energy difference can depend on the measurement outcomes as $\Delta F[Y_N] \equiv F(\lambda_{\mathrm{fin}}(Y_N)) - F(\lambda_{\mathrm{int}})$.

We have thus shown that the detailed fluctuation theorem still holds in the presence of feedback control if the measurement outcome is fixed.

Before proceeding to the main results of this chapter, we list our notations in Table 9.1.

9.3 Nonequilibrium Equalities with Feedback Control

We now derive the main results of this chapter. We derive the two types of the generalized nonequilibrium equalities with feedback control in Sects. 9.3.1 and 9.3.2, respectively. In particular, one generalization involves the mutual information, while the other involves the efficacy of feedback control.

9.3.1 Generalized Fluctuation Theorem with Mutual Information

To derive a generalized detailed fluctuation theorem, we first formulation the relevant backward probabilities. We consider the following type of "backward probability distribution":

$$P^\dagger[X_N^\dagger, Y_N] \equiv P[X_N^\dagger | \Lambda_N(Y_{N-1})^\dagger] P[Y_N], \tag{9.36}$$

which satisfies

$$\int P^\dagger[X_N^\dagger, Y_N] dX_N^\dagger dY_N = 1. \tag{9.37}$$

Definition (9.36) has a clear operational meaning. Suppose that we perform a forward experiment with feedback and obtain outcome Y_N. We then perform a backward experiment with protocol $\Lambda_N(Y_{N-1})^\dagger$. We repeat this set of the forward and backward experiments many times, and count the ratios of (X_N, Y_N) and $(X_N^\dagger, Y_N)$, which respectively give $P[X_N, Y_N]$ and $P^\dagger[X_N^\dagger, Y_N]$.

Noting Eq. (9.20) and the definition of the mutual information (9.28), we obtain a generalized detailed fluctuation theorem with feedback control [2]:

$$\frac{P^\dagger[X_N^\dagger, Y_N]}{P[X_N, Y_N]} = \exp\left(-\sigma[X_N, \Lambda_N(Y_{N-1})] - I_c[X_N : Y_N]\right), \tag{9.38}$$

Table 9.1 Symbols and their meanings

x	Phase-space point of the system
x^*	Time-reversal of x
$t_n \equiv n\Delta t$	Discretization of time
x_n	Phase-space point at time t_n
X_n	Trajectory of the phase-space point from time 0 to t_n, i.e. $X_n \equiv (x_0, x_1, \ldots, x_n)$
$X_n^\dagger$	Time-reversal of X_n, i.e. $X_n^\dagger \equiv (x_0^\dagger, x_1^\dagger, \ldots, x_n^\dagger)$ with $x_n^\dagger \equiv x_{N-n}^*$
λ_n	Controllable external parameters at time t_n
Λ_n	Control protocol from time 0 to t_n
$\Lambda_n^\dagger$	Backward control protocol of Λ_n, i.e. $\Lambda_n^\dagger \equiv (\lambda_0^\dagger, \lambda_1^\dagger, \ldots, \lambda_n^\dagger)$ with $\lambda_n^\dagger \equiv \lambda_{N-n-1}^*$.
$\sigma[X_n, \Lambda_n]$	Entropy production
y_n	Measurement outcome at time t_n
Y_n	Measurement outcomes from time 0 to t_n
$Y_n^\dagger$	Time-reversal of Y_n, i.e. $(y_{n_1}^\dagger, y_{n_2}^\dagger, \ldots, y_{n_{M_n}}^\dagger)$ with $y_{n_k}^\dagger \equiv y_{N-n_k}^*$, where M_n is the maximum integer satisfying $n_{M_n} \le n$
$\Lambda_n(Y_{n-1})$	Protocol of feedback control with outcomes Y_{n-1}
$P[X_n \vert \Lambda_n(Y_{n-1})]$	Probability of trajectory X_n under the condition that the protocol is given by $\Lambda_n(Y_{n-1})$ with Y_{n-1} being fixed outcomes
$\Lambda_n(Y_{n-1})^\dagger$	Time-reversal of $\Lambda_n(Y_{n-1})$: $\Lambda_n(Y_{n-1})^\dagger \equiv (\lambda_{N-1}(Y_{N-1})^*, \ldots, \lambda_{N-n-1}(Y_{N-n-1})^*)$
$P[y_n \vert X_n]$	Probability density of obtaining y_n under the condition of X_n, which characterizes the measurement error
$P_c[Y_n \vert X_n]$	$\equiv \prod_{k=1}^{M_n} P[y_{n_k} \vert X_{n_k}]$
$P[X_n, Y_n]$	Joint distribution of X_n and Y_n which is given by $P[X_n, Y_n] = P_c[Y_n \vert X_n] P[X_n \vert \Lambda_n(Y_{n-1})]$
$I[y_n : X_n \vert Y_{n-1}]$	Conditional mutual information obtained at time t_n under the condition that we have outcomes Y_{n-1}
$I_c[X_n : Y_n]$	Sum of the conditional mutual information: $I_c[X_n : Y_n] \equiv \prod_{k=1}^{M_n} I[y_{n_k} : X_{n_k} \vert Y_{n_k - 1}]$

where the effect of feedback control is involved by the term of the mutual information that is obtained in the forward experiments. We stress that, to obtain Eq. (9.38), we do not perform feedback control in the backward experiments. We just reverse forward protocol as Eq. (9.32) in the backward experiments. The same result for a special case was obtained in Ref. [20].

From the generalized detailed fluctuation theorem (9.38), we obtain a generalized integral fluctuation theorem [1, 2]:

$$\langle e^{-\sigma - I_{\mathrm{c}}} \rangle = 1. \tag{9.39}$$

Due to the concavity of the exponential function, we obtain a generalized second law of thermodynamics [1, 2, 37]

$$\langle \sigma \rangle \geq -\langle I_{\mathrm{c}} \rangle, \tag{9.40}$$

which means that the entropy production can be negative due to the effect of feedback control (or due to the action of Maxwell's demon), and that the lower bound of the entropy production is bounded by the mutual information $\langle I_{\mathrm{c}} \rangle$. We also obtain, by taking the ensemble average of the logarithm of the both sides of Eq. (9.38), that

$$\langle \sigma \rangle + \langle I_{\mathrm{c}} \rangle = \int dX_N dY_N P[X_N, Y_N] \ln \frac{P[X_N, Y_N]}{P^\dagger[X_N^\dagger, Y_N]}, \tag{9.41}$$

which is a generalization of the Kawai–Parrondo–Broeck (KPB) equality (8.18). We note that the right-hand side of Eq. (9.41) is positive because it is the Kullback–Leibler divergence between two probability distributions $P[X_N, Y_N]$ and $P^\dagger[X_N^\dagger, Y_N]$, which reproduces inequality (9.40). We note that equality in (9.40) is achieved if and only if $\sigma + I_{\mathrm{c}}$ does not fluctuate, or equivalently, $P[X_N, Y_N] = P^\dagger[X_N^\dagger, Y_N]$ holds.

If the joint distribution of σ and I_{c} is Gaussian, we have a generalized fluctuation-dissipation theorem from the second cumulant of Eq. (9.39):

$$\langle \sigma + I_{\mathrm{c}} \rangle = \frac{1}{2}(\langle (\sigma + I_{\mathrm{c}})^2 \rangle - \langle \sigma + I_{\mathrm{c}} \rangle^2), \tag{9.42}$$

in which there is a trade-off relation between the entropy production and the mutual information.

For the case in which $\sigma = \beta(W - \Delta F)$, Eq. (9.38) leads to a generalized Jarzynski equality:

$$\langle e^{-\beta(W - \Delta F) - I_{\mathrm{c}}} \rangle = 1, \tag{9.43}$$

and inequality (9.40) leads to

$$\langle W \rangle \geq \Delta F - \langle I_{\mathrm{c}} \rangle. \tag{9.44}$$

We note that Eq. (9.43) and inequality (9.44) are the generalizations of the results obtained in Refs. [1, 20]. By defining $\langle W_{\rm ext}\rangle \equiv -\langle W\rangle$ and setting $\Delta F = 0$, inequality (9.44) reduces to

$$\langle W_{\rm ext}\rangle \leq \langle I_{\rm c}\rangle, \tag{9.45}$$

which implies that we can extract a positive work up to the term that is proportional to the mutual information, from a thermodynamic cycle with a single heat bath with the assistance of feedback control or Maxwell's demon. The mutual information can be used as a "resource" of the work or the free energy. In the case of the Szilard engine, $\langle I_{\rm c}\rangle = \langle I\rangle = \ln 2$ and $\langle W_{\rm ext}\rangle = k_{\rm B}T\ln 2$ hold, and therefore the equality in inequality (9.45) is achieved. In fact, in the case of the Szilard engine, $\sigma + I = \beta(W - \Delta F) + I$ does not fluctuate, which is zero for both outcomes "left" and "right".

We note that, to obtain Eq. (9.39) or (9.43) experimentally or numerically, the condition of $P_{\rm c}[Y_N|X_N] \neq 0$ needs to be satisfied for all (X_N, Y_N). To explicitly see this, we write $P_{\rm c}[Y_N|X_N] \equiv \varepsilon > 0$. We then obtain

$$P[X_N, Y_N]e^{-\sigma - I_{\rm c}} = \varepsilon P[X_N|\Lambda_N(Y_{N-1})]\cdot\frac{1}{\varepsilon}e^{-\sigma + \ln P[Y_N]}, \tag{9.46}$$

which does not converge to zero with the limit of $\varepsilon \to 0$. On the other hand, in real experiments or numerical simulations, the events with $P[X_N, Y_N] = 0$ never occurs. Therefore, if $P_{\rm c}[Y_N|X_N] = 0$ holds for some (X_N, Y_N), the terms associated with zero-probability events make non-zero contributions to Eqs. (9.39) and (9.43); in such cases, we cannot obtain Eqs. (9.39) or (9.43) experimentally or numerically. On the contrary,

$$P[X_N, Y_N]I_{\rm c}^n = \varepsilon P[X_N|\Lambda_N(Y_{N-1})]\cdot\left(\ln\frac{\varepsilon}{P[Y_N]}\right)^n \tag{9.47}$$

converges to zero for all $n = 1, 2, \ldots$, in the limit of $\varepsilon \to 0$. Therefore, we can obtain $\langle I_{\rm c}^n\rangle$ experimentally and numerically even if $P_{\rm c}[Y_N|X_N] = 0$ for some (X_N, Y_N), and also obtain Eqs. (9.41), (9.42), and inequalities (9.40), (9.44), (9.45).

9.3.2 Generalized Fluctuation Theorem with Efficacy Parameter

We next derive a different type of generalization of nonequilibrium equalities. In this subsection, we assume that the measurements are Markovian (i.e., $P[y_n|X_n] = P[y_n|x_n]$ holds).

We perform forward experiments with measurements at times $t_{n_1}, t_{n_2}, \ldots, t_{n_M}$ with feedback control, and perform backward experiments without feedback but only with measurements at times $t_{N-n_M}, t_{N-n_{M-1}}, \ldots, t_{N-n_1}$.

Let $Y'_N \equiv (y'_{N-n_M}, y'_{N-n_{M-1}}, \ldots, y'_{N-n_1})$ be the measurement outcomes in the backward measurements. Then the probability of obtaining Y'_N under the condition of $X_N^\dagger$ is given by

$$P_{\mathrm{c}}[Y'_N|X_N^\dagger] \equiv \prod_{k=1}^{M} P[y'_{N-n_k}|x_{N-n_k}^\dagger]. \tag{9.48}$$

Therefore the probability of obtaining Y'_N under protocol $\Lambda(Y_N)^\dagger$ is

$$P[Y'_N|\Lambda_N(Y_{N-1})^\dagger] = \int P_{\mathrm{c}}[Y'_N|X_N^\dagger]P[X_N^\dagger|\Lambda_N(Y_{N-1})^\dagger]dX_N^\dagger, \tag{9.49}$$

which is normalized as

$$\int P[Y'_N|\Lambda_N(Y_{N-1})^\dagger]dY'_N = 1, \tag{9.50}$$

where the probability variable Y'_N is independent of Y_N.

We then consider the time-reversed sequence of Y_N. Let y_n^* be the time-reversal of y_n; for example, if we measure the momentum, then $y_n^* = -y_n$. We write $Y_N^\dagger \equiv (y_{N-n_M}^*, y_{N-n_{M-1}}^*, \dots, y_{N-n_1}^*)$. The probability of $Y'_N = Y_N^\dagger$ in the backward experiments is given by

$$P[Y_N^\dagger|\Lambda_N(Y_{N-1})^\dagger] = \int P_{\mathrm{c}}[Y_N^\dagger|X_N^\dagger]P[X_N^\dagger|\Lambda_N(Y_{N-1})^\dagger]dX_N^\dagger, \tag{9.51}$$

which is the probability of obtaining the time-reversed outcomes by time-reversed measurements during the time-reversed protocol. We stress that

$$\int P[Y_N^\dagger|\Lambda_N(Y_{N-1})^\dagger]dY_N^\dagger \neq 1 \tag{9.52}$$

in general because $Y_N^\dagger$ is no longer independent of Y_{N-1}.

In the following, we assume that the measurements have the time-reversed symmetry

$$P[y_n^*|x_n^*] = P[y_n|x_n]. \tag{9.53}$$

for all n, which leads to

$$P_{\mathrm{c}}[Y_n^\dagger|X_n^\dagger] = P_{\mathrm{c}}[Y_n|X_n]. \tag{9.54}$$

We then have the "renormalized" (or "coarse-grained") detailed fluctuation theorem [1, 2, 38]

$$\frac{P[Y_N^\dagger|\Lambda_N(Y_{N-1})^\dagger]}{P[Y_N]} = e^{-\sigma'[Y_N]}, \tag{9.55}$$

where $\sigma'[Y_N]$ is the "renormalized" (or "coarse-grained") entropy production defined as

$$\sigma'[Y_N] \equiv -\ln\langle e^{-\sigma}\rangle_{Y_N}$$
$$= -\ln \int dX_N e^{-\sigma[X_N, \Lambda_N(Y_{N-1})]} P[X_N|Y_N]. \tag{9.56}$$

Equality (9.55) implies that the detailed fluctuation theorem dose not change its form under the coarse-graining, if we correspondingly define the appropriate coarse-grained entropy production. From the concavity of the exponential function, we obtain $\sigma'[Y_N] \leq \langle\sigma\rangle_{Y_N}$ and $\langle\sigma'\rangle \leq \langle\sigma\rangle$. The same result for a different setup has been obtained in Ref. [38].

The proof of (9.55) goes as follows. From the definition of $\sigma'[Y_N]$ and the detailed fluctuation theorem (9.33), we have

$$\begin{aligned} e^{-\sigma'[Y_N]} &= \int dX_N \frac{P[X_N^\dagger|\Lambda_N(Y_{N-1})^\dagger]}{P[X_N|\Lambda_N(Y_{N-1})]} P[X_N|Y_N] \\ &= \int dX_N \frac{P[X_N^\dagger|\Lambda_N(Y_{N-1})^\dagger]}{P[X_N|\Lambda_N(Y_{N-1})]} \frac{P[X_N, Y_N]}{P[Y_N]} \\ &= \frac{1}{P[Y_N]} \int dX_N P[X_N^\dagger|\Lambda_N(Y_{N-1})^\dagger] P_c[Y_N|X_N] \\ &= \frac{1}{P[Y_N]} \int dX_N P[X_N^\dagger|\Lambda_N(Y_{N-1})^\dagger] P_c[Y_N^\dagger|X_N^\dagger]. \end{aligned} \tag{9.57}$$

In the last line, we used the time-reversal symmetry (9.54) of the measurements. By noting Eq. (9.51), we obtain (9.55). We note that Eq. (9.55) holds regardless of the presence of feedback control. Without feedback control, Eq. (9.55) reduces to

$$\frac{P^\dagger[Y_N^\dagger]}{P[Y_N]} = e^{-\sigma'[Y_N]}. \tag{9.58}$$

By taking the ensemble average of both sides of Eq. (9.55) and noting that $\langle e^{-\sigma'}\rangle = \langle e^{-\sigma}\rangle$ holds, we obtain the second generalization of the integral fluctuation theorem [2]

$$\langle e^{-\sigma}\rangle = \gamma, \tag{9.59}$$

where γ is the efficacy parameter of feedback control defined as

$$\gamma \equiv \int P[Y_N^\dagger|\Lambda_N(Y_{N-1})^\dagger] dY_N^\dagger, \tag{9.60}$$

which is the sum of probabilities of obtaining the time-reversed outcomes by the time-reversed measurements during the time-reversed protocols (see Fig. 9.2). Here γ characterizes a kind of reversibility of the feedback protocol: the more reversible the control is, the more the amount of γ is. Without feedback control, γ reduces to 1 as

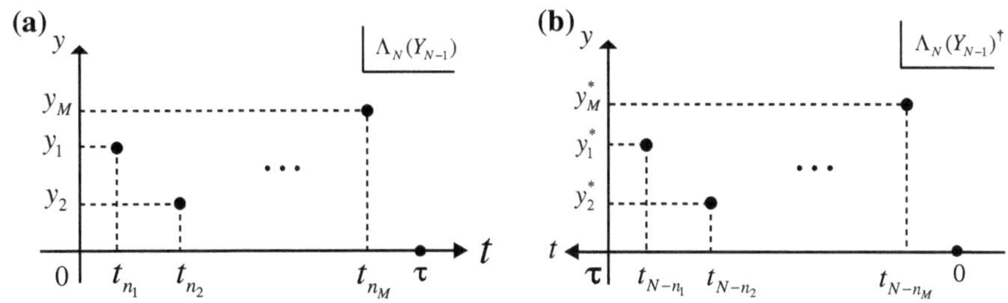

Fig. 9.2 Forward and backward measurement outcomes (reproduced from Ref. [2] with permission). **a** Forward outcomes Y_N with forward protocol $\Lambda_N(Y_{N-1})$. **b** Backward outcomes $Y_N^\dagger$ with backward protocol $(\Lambda_N(Y_{N-1})^\dagger)$

$$\gamma \equiv \int P[Y_N^\dagger]dY_N^\dagger = 1, \tag{9.61}$$

which vindicates the original integral fluctuation theorem. If $\sigma = \beta(W - \Delta F)$ holds, Eq. (9.59) leads to the second generalization of the Jarzynski equality [1, 2]:

$$\langle e^{-\beta(W-\Delta F)} \rangle = \gamma. \tag{9.62}$$

In the case of the Szilard engine, $\gamma = 2$ holds corresponding to $W = -k_B T \ln 2$ and $\Delta F = 0$ [1, 2].

We stress that σ and γ can be measured independently, because σ is obtained from the forward experiments with feedback and γ is obtained from the backward experiments without feedback. Therefore Eqs. (9.59) and (9.62) can be directly verified in experiments. In fact, Eq. (9.62) has been verified in a real experiment by using a feedback-controlled ratchet with a Brownian particle [21].

From Eq. (9.38), we have the second generalization of the second law of thermodynamics

$$\langle \sigma \rangle \geq -\ln \gamma. \tag{9.63}$$

The equality in inequality (9.63) is achieved if σ does not fluctuate. We note that, if the distribution of σ is Gaussian, we have a generalized fluctuation-dissipation theorem

$$\langle \sigma \rangle + \ln \gamma = \frac{1}{2}(\langle \sigma^2 \rangle - \langle \sigma \rangle^2). \tag{9.64}$$

While the first generalization (9.39) only involves the term of the obtained information, the second generalization (9.59) involves the term of feedback efficacy. To understand the relationship between the mutual information I_c and the feedback efficacy γ, we introduce the notation

$$C[A] \equiv -\ln\langle e^{-A} \rangle \tag{9.65}$$

for any probability variable A. We note that, if A can be written as $A = tA'$ with t being a real number and A' being another probability variable, then $C[A]$ is the cumulant generation function of A'. By using this notation, we have

$$C[\sigma] + C[I_c] - C[\sigma + I_c] = -\ln \gamma, \tag{9.66}$$

because $C[\sigma] = -\ln \gamma$ holds as in Eq. (9.59), $C[I_c] = 0$ holds as Eq. (9.29), and $C[\sigma + I_c] = 0$ holds as Eq. (9.39). Equality (9.66) implies that $-\ln \gamma$ is a measure of the correlation between σ and I_c. This can be more clearly seen by the cumulant expansion of Eq. (9.66) if the joint distribution of σ and I_c is Gaussian:

$$\langle \sigma I_c \rangle - \langle \sigma \rangle \langle I_c \rangle = -\ln \gamma. \tag{9.67}$$

Therefore, γ characterizes how efficiently we use the obtained information to decrease the entropy production by feedback control: if γ is large, the more I_c we obtain the less σ is.

We can also derive another nonequilibrium equality which also gives us the information about the feedback efficacy. By taking logarithm of the both sides of Eq. (9.55), we obtain

$$\langle \sigma' \rangle = \int dY_N P[Y_N] \ln \frac{P[Y_N]}{P[Y_N^\dagger | \Lambda_N(Y_{N-1})^\dagger]}, \tag{9.68}$$

which is a generalization of Eq. (8.18). The same result in a different situation has also been obtained in Ref. [38]. Equality (9.68) implies that the renormalized entropy production equals the Kullback–Leibler divergence-like quantity between the forward probability $P[Y_N]$ and the backward probability $P[Y_N^\dagger | \Lambda_N(Y_{N-1})^\dagger]$. In fact, without feedback control, the right-hand side of Eq. (9.68) reduces to the Kullback–Leibler divergence between $P[Y_N]$ and $P^\dagger[Y_N^\dagger]$ and therefore the both sides of Eq. (9.68) are positive in consistent with the second law of thermodynamics. On the contrary, in the presence of feedback control, the right-hand side is no longer the Kullback–Leibler divergence, because $P[Y_N^\dagger | \Lambda_N(Y_{N-1})^\dagger]$ is not a normalized probability distribution in terms of $Y_N^\dagger$. Therefore the both sides of (9.68) can be negative. Since $\langle \sigma' \rangle \leq \langle \sigma \rangle$, the entropy production $\langle \sigma \rangle$ is bounded by the right-hand side of Eq. (9.68):

$$\langle \sigma \rangle \geq \int dY_N P[Y_N] \ln \frac{P[Y_N]}{P[Y_N^\dagger | \Lambda_N(Y_{N-1})^\dagger]}. \tag{9.69}$$

Without feedback control, the right-hand side of (9.69) gives a positive bound, while, with feedback control, the right-hand side can give a negative bound. We note that, for a quantum generalization of the Szilard engine with multi-particles, essentially the same result as Eq. (9.68) has been obtained [39].

We note that specific versions of our results in this chapter were obtained elsewhere. We have derived two types of the generalized Jarzynski equality for the cases with a single measurement in the presence of a single heat bath in Ref. [1], and for the cases with multi-measurements and multi-heat baths in Ref. [2]. In Ref. [20], the detailed fluctuation theorem and the Jarzynski equality were obtained for the cases with multi-measurements and feedback in the presence of a single heat bath. In Ref. [19], a generalized Jarzynski equality was also obtained for the Kalman filter and the optimal control. The results in this chapter include all of the above results, and generalize them to the cases of multi-heat baths and non-Markovian measurements.

We also note that a generalized Jarzynski equality (9.62) with a single measurement was experimentally verified by using a feedback-controlled ratchet with a colloidal particle [21]. Moreover, Eq. (9.62) has been generalized to quantum systems [22].

9.4 Examples

We next discuss two examples which illustrate the essential features of our general results. We consider a generalized Szilard engine analytically in Sect. 9.4.1, and a feedback-controlled ratchet numerically in Sect. 9.4.2.

9.4.1 Szilard Engine with Measurement Errors

We also check the generalized Jarzynski equalities in the model discussed in Sect. 6.3 for arbitrary v_0, v_1, and ε [2].

We first note that $I[x:y]$ is given by $\ln 2(1-\varepsilon)$ when $(x, y) = (0, 0)$, $\ln 2\varepsilon$ when $(x, y) = (0, 1)$, $\ln 2\varepsilon$ when $(x, y) = (1, 0)$, and $\ln 2(1-\varepsilon)$ when $(x, y) = (1, 1)$. Therefore we obtain

$$\langle e^{-\beta W - I} \rangle = \frac{v_0 + (1 - v_0) + (1 - v_1) + v_1}{2} = 1, \tag{9.70}$$

which confirms Eq. (9.43).

We next consider the second generalization (9.62) of the Jarzynski equality. Corresponding to two measurement outcomes $y = 0, 1$, we have two backward control protocols as follows (see also Fig. 9.3).

Step 1. Initial state. The initial state of the backward control is in the thermal equilibrium.

Step 2. Insertion of the barrier. Corresponding to Step 5 of the forward process, we insert the barrier and decide the box into two boxes, because the time-reversal of the barrier removal is the barrier insertion. Corresponding to $y = 0$ or $y = 1$ in the forward process, we divide the box with the ratio $v_0 : 1 - v_0$ or $1 - v_1 : v_1$, respectively.

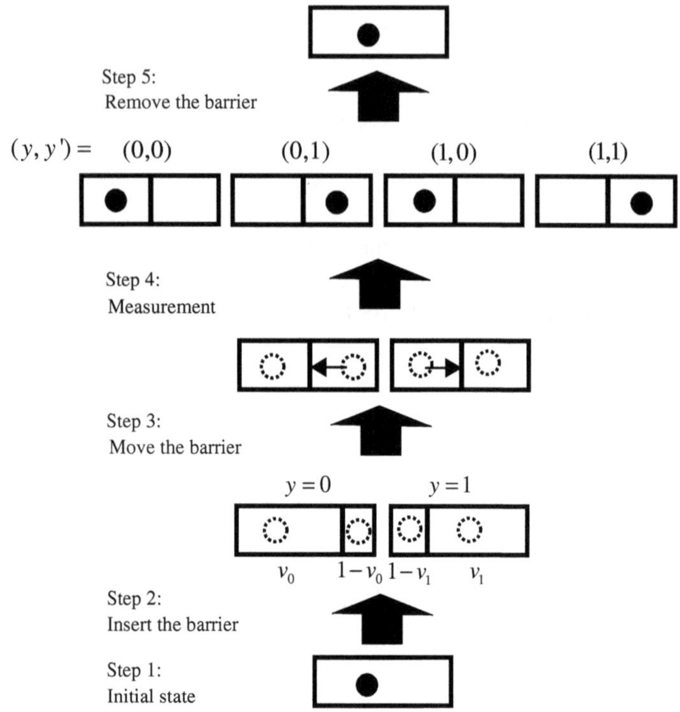

Fig. 9.3 Backward processes of the generalized Szilard engine (reproduced from Ref. [2] with permission). Corresponding to y that denotes the measurement outcomes in the forward process, we have two control protocols in the backward process, where y' denotes the measurement outcomes in the backward process

Step 3. Moving the barrier. We next move the barrier to the middle of the box quasi-statically and isothermally. This is the time-reversal of the feedback control in Step 4 of the forward process.

Step 4. Measurement. We perform the measurement to find in which box the particle is in. Corresponding to the backward protocol with $y = 0$, we obtain the outcomes of backward measurement $y' = 0$ with probability $P[y' = 0|\Lambda(y = 0)^\dagger] = v_0(1-\varepsilon) + (1-v_0)\varepsilon$ and $y' = 1$ with probability $P[y' = 1|\Lambda(y = 0)^\dagger] = v_0\varepsilon + (1-v_0)(1-\varepsilon)$. On the other hand, corresponding to the backward protocol with $y = 1$, we obtain the outcomes of backward measurement $y' = 0$ with probability $P[y' = 0|\Lambda(y = 1)^\dagger] = v_1\varepsilon + (1-v_1)(1-\varepsilon)$ and $y' = 1$ with probability $P[y' = 1|\Lambda(y = 1)^\dagger] = v_1(1-\varepsilon) + (1-v_0)\varepsilon$.

Step 5. Removal of the barrier. We remove the barrier and the system goes back to the initial state. This is the time-reversal of the barrier insertion in Step 2 of the forward process.

From Step 4 of the backward process, we have

$$\gamma \equiv P[y' = 0|\Lambda(y = 0)^{\dagger}] + P[y' = 1|\Lambda(y = 1)^{\dagger}] \\ = (1 - \varepsilon)(v_0 + v_1) + \varepsilon(2 - v_0 - v_1). \tag{9.71}$$

On the other hand, we can straightforwardly obtain

$$\langle e^{-\beta W} \rangle = (1 - \varepsilon)(v_0 + v_1) + \varepsilon(2 - v_0 - v_1), \tag{9.72}$$

which confirms Eq. (9.62).

9.4.2 Feedback-Controlled Ratchet

We next discuss a model of Brownian motors [40–45], in particular feedback-controlled ratchet [2, 11, 14, 16]. We consider a rotating Brownian particle with a periodic boundary condition. As in Eq. (8.61), we assume that the particle obeys the overdamped Langevin equation:

$$\eta \frac{dx(t)}{dt} = -\frac{\partial V(x, \lambda)}{\partial x} + \sqrt{2\eta k_{\rm B} T}\xi(t), \tag{9.73}$$

where x is the position or the angle of the particle, and its boundary condition is assumed to be given by $x = x + L$ with L being a constant. In the following, we restrict the particle's position to $-L/2 \leq x < L/2$. We assume that control parameter λ takes two values ($\lambda = 0$ or 1), and corresponding to it, the ratchet potential V takes the following two profiles (Fig. 9.4):

$$V(x, 0) = \begin{cases} K(x + L/2)/l \\ (-L/2 \leq x < -L/2 + l), \\ -K(x - L/2)/(L - l) \\ (-L/2 + l \leq x < L), \end{cases} \tag{9.74}$$

$$V(x, 1) = \begin{cases} -K(x + L/2 - l)/(L - l) \\ (-L/2 \leq x < -L/2 + l), \\ K(x + L/2 + 2l)/l \\ (-L/2 + l \leq x < -L/2 + 2l), \\ -K(x - L/2 - l)/(L - l) \\ (-L/2 + 2l \leq x < L/2), \end{cases} \tag{9.75}$$

where l is a constant with $0 < l < L/2$, and K is a positive constant that characterizes the height of the potential.

We start with the initial equilibrium with parameter $\lambda = 0$, and control the system from time $t = 0$ to τ with the following three protocols [2].

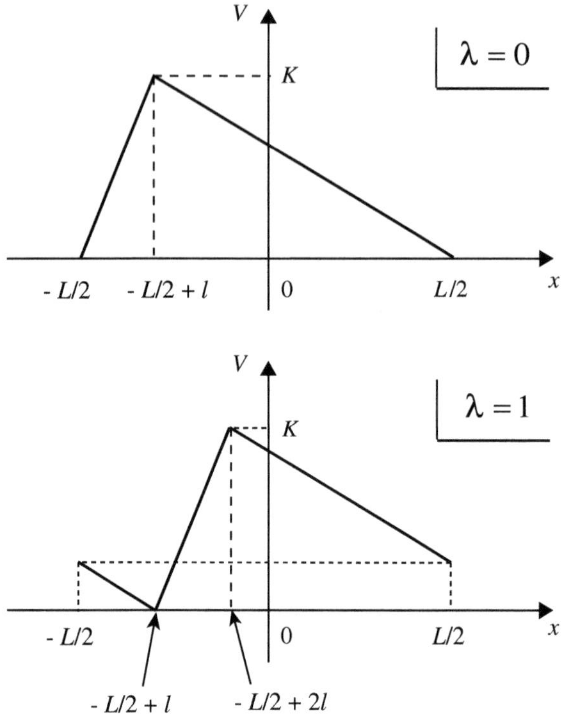

Fig. 9.4 Two shapes of potential $V(x, \lambda)$ corresponding to $\lambda = 0, 1$ (reproduced from Ref. [2] with permission)

1. *Trivial control.* We do not change the parameter $\lambda = 0$.
2. *Flashing ratchet.* At times $t = m\tau_0$ with m being integers and τ_0 being a constant, we switch parameter λ from 0 to 1 or from 1 to 0 periodically.
3. *Feedback-controlled ratchet.* At times $t = m\tau_0$, we switch the parameter with the following feedback protocol. We measure the position x at $t = m\tau_0$ without error. We then set$\lambda = 1$ from $t = m\tau_0$ to $(m+1)\tau_0$ if and only if the outcome is in $-L/2 \leq x < -L/2 + l$. Otherwise, parameter λ is set to 0.

For numerical simulations, we set $l = 3L/10$, $K = 3k_{\mathrm{B}}T$, $\tau_0 = 0.05$, and $\tau = 0.25$, with units $k_{\mathrm{B}}T = 1$, $L = 1$, and $\eta/2 = 1$. We performed the simulations by discretizing Eq. (9.73) with $\Delta t = 0.00025$ for 1,000,000 samples. We note that, to obtain the initial thermal equilibrium, we waited $\tau_{\mathrm{wait}} = 0.5$ and checked that the system was fully thermalized in the periodic ratchet with parameter $\lambda = 0$.

The time evolution of the ensemble average $\langle x(t) \rangle$ is plotted in Fig. 9.5a for the above three protocols [2]. As expected, nothing happens with the first protocol, while the particle is transported to the right on average in the second and third protocols. In the case of the feedback-controlled ratchet, the particle is transported to the right faster than the case of the flashing ratchet. Figure 9.5b shows the time evolution of the work $\langle W(t) \rangle$ that is inputted to the particle. The work is induced only in the

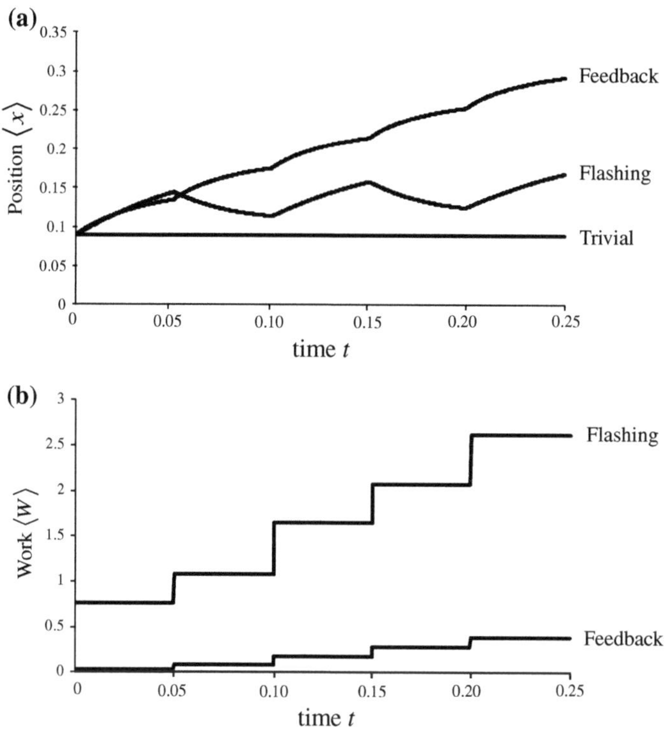

Fig. 9.5 Numerical results on the ratchet models (reproduced from Ref. [2] with permission). **a** Numerical results of the ensemble average of trajectory $x(t)$ corresponding to the three control protocols: the trivial control, the flashing ratchet, and the feedback-controlled ratchet. **b** Numerical result of the ensemble average of the work $W(t)$ corresponding to the flashing ratchet and the feedback-controlled ratchet

switching times. We find that, in order to transport the particle, the energy input to the particle with feedback control is smaller than that with the flashing.

Figure 9.6 shows the left-hand side of the Jarzynski equality $\langle e^{-\beta(W-\Delta F)}\rangle$ for the flashing and feedback-controlled ratchet, and the efficacy parameter γ for the feedback-controlled ratchet [2]. We note that $\Delta F = 0$ always holds. With feedback control, $\langle e^{-\beta(W-\Delta F)}\rangle$ increases from 1 as the number of switchings increases, while, without feedback control, $\langle e^{-\beta W}\rangle$ converges to 1 for all switching times in consistent with the original Jarzynski equality. On the other hand, to obtain γ, we numerically performed the backward experiments. The discretization of the time is $\Delta t = 0.0005$, and the number of the samples is 10,000 samples for each trajectory of $\lambda(t)$. We note that the number of the trajectories of λ is given by 2^m with m times of switchings. Figure 9.6 shows a good coincidence between $\langle e^{-\beta W}\rangle$ and γ, which confirms the validity of Eq. (9.62) in the feedback-controlled ratchet.

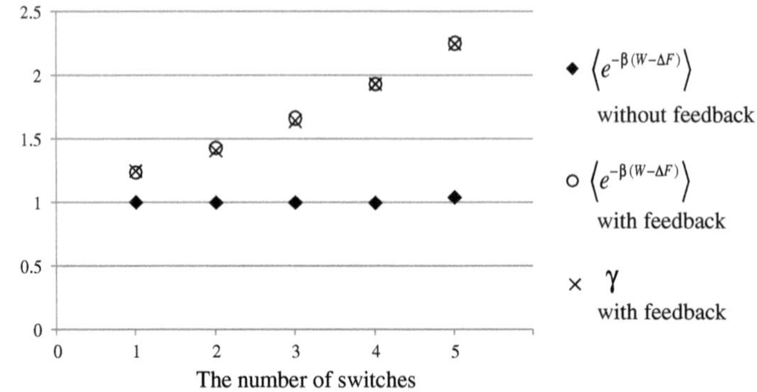

Fig. 9.6 Numerical tests of the Jarzynski equality for the flashing ratchet and a generalized Jarzynski equality (9.62) for the feedback-controlled ratchet (reproduced from Ref. [2] with permission)

References

1. T. Sagawa, M. Ueda, Phys. Rev. Lett. **104**, 090602 (2010)
2. T. Sagawa, M. Ueda, Phys. Rev. E **85**, 021104 (2012)
3. J.C. Maxwell, *Theory of Heat* (Appleton, London, 1871)
4. H.S. Leff, A.F. Rex (eds.), *Maxwell's Demon 2: Entropy, Classical and Quantum Information, Computing* (Princeton University Press, New Jersey, 2003)
5. K. Maruyama, F. Nori, V. Vedral, Rev. Mod. Phys. **81**, 1 (2009)
6. O.J.E. Maroney, in Information Processing and Thermodynamic Entropy, ed. by E.N. Zalta. The Stanford Encyclopedia of Philosophy (Fall 2009 Edition).
7. T. Sagawa, M. Ueda, Information Thermodynamics: Maxwell's Demon in Nonequilibrium Dynamics, arXiv:1111.5769 (2011); to appear in: R. Klages, W. Just, C. Jarzynski (eds.), Nonequilibrium Statistical Physics of Small Systems: Fluctuation Relations and Beyond (Wiley-VCH, Weinheim, 2012).
8. C.M. Caves, Phys. Rev. Lett. **64**, 2111 (1990)
9. H. Touchette, S. Lloyd, Phys. Rev. Lett. **84**, 1156 (2000)
10. H. Touchette, S. Lloyd, Phys. A **331**, 140 (2004)
11. F.J. Cao, L. Dinis, J.M.R. Parrondo, Phys. Rev. Lett. **93**, 040603 (2004)
12. K.H. Kim, H. Qian, Phys. Rev. Lett. **93**, 120602 (2004)
13. K.H. Kim, H. Qian, Phys. Rev. E **75**, 022102 (2007)
14. B.J. Lopez, N.J. Kuwada, E.M. Craig, B.R. Long, H. Linke, Phys. Rev. Lett. **101**, 220601 (2008)
15. F.J. Cao, M. Feito, Phys. Rev. E **79**, 041118 (2009)
16. M. Feito, J.P. Baltanas, F.J. Cao, Phys. Rev. E **80**, 031128 (2009)
17. F.J. Cao, M. Feito, H. Touchette, Phys. A **388**, 113 (2009)
18. M. Ponmurugan, Phys. Rev. E **82**, 031129 (2010)
19. Y. Fujitani, H. Suzuki, J. Phys. Soc. Jpn. **79**, 104003 (2010)
20. J.M. Horowitz, S. Vaikuntanathan, Phys. Rev. E **82**, 061120 (2010)
21. S. Toyabe, T. Sagawa, M. Ueda, E. Muneyuki, M. Sano, Nat. Phys. **6**, 988 (2010)
22. Y. Morikuni, H. Tasaki, J. Stat. Phys. **143**, 1 (2011)
23. S. Ito, M. Sano, Phys. Rev. E **84**, 021123 (2011)
24. J.M. Horowitz, J.M.R. Parrondo, Europhys. Lett. **95**, 10005 (2011)
25. D. Abreu, U. Seifert, Europhys. Lett. **94**, 10001 (2011)
26. S. Vaikuntanathan, C. Jarzynski, Phys. Rev. E **83**, 061120 (2011)
27. T. Sagawa, J. Phys. Conf. Ser. **297**, 012015 (2011)

28. D.V. Averin, M. Möttönen, J.P. Pekola, Phys. Rev. B **84**, 245448 (2011)
29. J.M. Horowitz, J.M.R. Parrondo, New J. Phys. **13**, 123019 (2011)
30. S. Lahiri, S. Rana, A.M. Jayannavar, J. Phys. A Math. Theor. **45**, 065002 (2012)
31. D. Abreu, U. Seifert, Phys. Rev. Lett. **108**, 030601 (2012)
32. T. Schreiber, Phys. Rev. Lett. **85**, 461 (2000)
33. J.C. Doyle, B.A. Francis, A.R. Tannenbaum, *Feedback Control Theory* (Macmillan, New York, 1992)
34. K.J. Åstrom, R.M. Murray, Feedback Systems: An Introduction for Scientists and Engineers (Princeton University Press, Princeton, 2008).
35. G. Welch, G. Bishop, *An introduction to the Kalman filter, Technical Report TR 95–041* (University of North Carolina, Department of Computer Science, 1995)
36. D.P. Bertsekas, *Dynamic Programming and Optimal Control* (Athena Scientific, Belmont, 2005)
37. T. Sagawa, M. Ueda, Phys. Rev. Lett. **100**, 080403 (2008)
38. R. Kawai, J.M.R. Parrondo, C. Van den Broeck, Phys. Rev. Lett. **98**, 080602 (2007)
39. S.W. Kim, T. Sagawa, S. De Liberato, M. Ueda, Phys. Rev. Lett. **106**, 070401 (2011)
40. R.D. Vale, F. Oosawa, Adv. Biophys. **26**, 97 (1990)
41. F. Julicher, A. Ajdari, J. Prost, Rev. Mod. Phys. **69**, 1269 (1997)
42. J.M.R. Parrondo, B.J. De Cisneros, Appl. Phys. A **75**, 179 (2002)
43. P. Reimann, Phys. Rept. **361**, 57 (2002)
44. P. Hänggi, F. Marchesoni, F. Nori, Ann. Phys. **14**, 51 (2005)
45. P. Hänggi, F. Marchesoni, Rev. Mod. Phys. **81**, 387 (2009)

Chapter 10
Conclusions

Modern nonequilibrium statistical mechanics provides derivations of the second law of thermodynamics based on the microscopic quantum dynamics of thermodynamic systems [1–4]. By combining the derivations with quantum information theory, generalizations of the second law of thermodynamics have been derived for quantum information processing such as quantum measurement, quantum feedback control, and information erasure [5–13]. Moreover, generalized nonequilibrium equalities such as the fluctuation theorem and the Jarzynski equality have been derived in the presence of feedback control [14–22]. These results are applicable to small thermodynamic systems that can be precisely controlled by modern experimental technologies. These results are also closely related to the fundamental problem of Maxwell's demon [23–27], which can be regarded as a feedback controller acting on thermodynamic systems.

The main results of this thesis are in Chaps. 6, 7 and 9, which will be summarized with several discussions in the followings.

In Chap. 5, we have discussed a possible derivation of the second law of thermodynamics based on quantum-statistical mechanics. We have formulated the total system of a thermodynamic system and heat baths obeys a unitary evolution, and the initial states of the heat baths are in the canonical distribution. As a result, we have derived a general form of the second law (5.41), which leads to several expressions of the second law such as Kelvin's principle and the Clausius inequality. The reason why we can derive the second law from the unitary dynamics lies in the fact that we have set the initial canonical distributions in which the von Neumann entropy takes the maximum values under a given amount of the energy. While this type of derivation of the second law is a standard method in modern statistical mechanics, there would be a room to make the proof physically clearer. In fact, it has been recognized that a thermal equilibrium state of the total system from the macroscopic point of view do not necessarily correspond to the canonical (or microcanonical) distribution from the microscopic point of view; in fact, even a pure state can behave as a thermal equilibrium state [28–36]. Therefore, it is a future challenge to relax the condition of

T. Sagawa, *Thermodynamics of Information Processing in Small Systems*,
Springer Theses, DOI: 10.1007/978-4-431-54168-4_10,

the initial canonical distribution. We note that some approaches have been studied in this direction [37–40].

In Chaps. 6 and 7, we have generalized the second law of thermodynamics to information processing processes by involving quantum information theory to the proof of the second law that is studied in Chap. 5.

In Chap. 6, we have derived the maximum work (6.17) that can be extracted from thermodynamic systems that are subject to feedback control [7, 13]. The maximum work is given by the term of the free energy and the term that is proportional to the QC-mutual information obtained by a measurement. The QC-mutual information, which has been discussed in Chap. 4 in detail, represents a kind of correlation between the measured quantum system and the measurement outcomes. The QC-mutual information reduces to the classical mutual information in the cases of classical measurements.

In Chap. 7, we have derived the minimal works (7.34) and (7.20) that are performed on memories during the measurement and information erasure [11, 13]. The memories can be regarded as the physical implementations of Maxwell's demon, and therefore our results have identified the minimal work that is needed for the demon to work. The minimal work for the erasure leads to Landauer's principle for special cases. Our results lead to the trade-off relation between the works that are needed for the measurement and erasure, and the lower bound (7.43) of the sum of the works is determined only by the temperature and the QC-mutual information.

The results obtained in Chaps. 6 and 7 implies that the excess work that is extracted from Maxwell's demon is compensated for by the total work that is needed for the measurement and the erasure. Therefore, our results enables us to reconcile the Maxwell's demon with the second law of thermodynamics, which is different from the previous approaches for the reconciliation. The derived inequalities in Chaps. 6 and 7 are the generalizations of the second law of thermodynamics, which can be applied to the information processing processes.

We note that, our results in Chap. 6 and Sect. 7.3 have been derived under the assumption that any element of the POVM corresponds to a single measurement operator so that $\hat{E}_k = \hat{M}_k^\dagger \hat{M}_k$. To relax this assumption is a future challenge.

In Chap. 9, we have generalized the nonequilibrium equalities such as the Jarzynski equality to the situations in which classical thermodynamic systems are subject to feedback control [14, 22]. In particular, we have derived two types of the generalized Jarzynski equalities (9.43) and (9.62): the former involves the term of the mutual information obtained by the measurements, and the latter involves the term of the efficacy of feedback control. Inequality (9.43) reproduces the result in Chap. 6 for the classical cases, and inequality (9.62) has been experimentally verified by using a colloidal particle. We have demonstrated the obtained equalities with a generalized Szilard engine by the analytic method and with a feedback controlled ratchet by the numerical method.

In our results, the thermodynamic quantities, such as the free energy and the work, and the information contents such as the Shannon information and the mutual information, are treated on an equal footing. Therefore, our theory developed in this thesis can be regarded as constituting "information thermodynamics." Information thermodynamics sheds fundamental lights on the foundations of thermodynamics and statistical mechanics through the paradox of Maxwell's demon, and has potential applications to information processing in small systems such as nanomachines and nanodevices.

References

1. J. Kurchan, arXiv:cond-mat/0007360 (2000).
2. H. Tasaki, arXiv:cond-mat/0009244 (2000).
3. C. Jarzynski, D.K. Wójcik, Phys. Rev. Lett. **92**, 230602 (2004)
4. T. Sagawa, Second law-like inequalities with quantum relative entropy: an introduction, in Lectures on Quantum Computing, Thermodynamics and Statistical Physics. Kinki University Series on Quantum Computing (World Scientific, Singapore, 2012); e-print: arXiv:1202.0983. (To appear).
5. S. Lloyd, Phys. Rev. A **39**, 5378 (1989)
6. M.A. Nielsen, C.M. Caves, B. Schumacher, H. Barnum, Proc. R. Soc. Lond. A **454**, 277 (1998)
7. T. Sagawa, M. Ueda, Phys. Rev. Lett. **100**, 080403 (2008)
8. A.E. Allahverdyan, T.M. Nieuwenhuizen, Phys. Rev. E **64**, 0561171 (2001)
9. C. Horhammer, H. Buttner, J. Stat. Phys. **133**, 1161 (2008)
10. O.J.E. Maroney, Phys. Rev. E **79**, 031105 (2009)
11. T. Sagawa, M. Ueda, Phys. Rev. Lett. **102**, 250602 (2009)
12. T. Sagawa, M. Ueda, Phys. Rev. Lett. 106, 189901(E) (2011).
13. T. Sagawa, Prog. Theor. Phys. **127**, 1 (2012)
14. T. Sagawa, M. Ueda, Phys. Rev. Lett. **104**, 090602 (2010)
15. M. Ponmurugan, Phys. Rev. E **82**, 031129 (2010)
16. Y. Fujitani, H. Suzuki, J. Phys. Soc. Jpn. **79**, 104003 (2010)
17. J.M. Horowitz, S. Vaikuntanathan, Phys. Rev. E **82**, 061120 (2010)
18. S. Toyabe, T. Sagawa, M. Ueda, E. Muneyuki, M. Sano, Nat. Phys. **6**, 988 (2010)
19. Y. Morikuni, H. Tasaki, J. Stat. Phys. **143**, 1 (2011)
20. S. Lahiri, S. Rana, A.M. Jayannavar, J. Phys. A Math. Theor. **45**, 065002 (2012)
21. D. Abreu, U. Seifert, Phys. Rev. Lett. **108**, 030601 (2012)
22. T. Sagawa, M. Ueda, Phys. Rev. E **85**, 021104 (2012)
23. J.C. Maxwell, *Theory of Heat* (Appleton, London, 1871)
24. H.S. Leff, A.F. Rex (eds.), *Maxwell's Demon 2: Entropy, Classical and Quantum Information, Computing* (Princeton University Press, Princeton, 2003)
25. K. Maruyama, F. Nori, V. Vedral, Rev. Mod. Phys. **81**, 1 (2009)
26. O.J.E. Maroney, Information processing and thermodynamic entropy, in The Stanford Encyclopedia of Philosophy, Fall 2009 edn., ed. by E.N. Zalta (Springer, Berlin, 2009).
27. T. Sagawa, M. Ueda, Information thermodynamics: Maxwell's Demon in nonequilibrium dynamics, in Nonequilibrium Statistical Physics of Small Systems: Fluctuation Relations and Beyond, ed. by R. Klages, W. Just, C. Jarzynski (Wiley-VCH, Weinheim, 2012). arXiv:1111.5769 (2011) (To appear).
28. J. von Neumann, Z. Phys. 57, 30 (1929) (Eng. Trans. in arXiv:1003.2133).
29. J.M. Deutsch, Phys. Rev. A **43**, 2046 (1991)
30. M. Srednicki, Phys. Rev. E **50**, 888 (1994)
31. H. Tasaki, Phys. Rev. Lett. **80**, 1373 (1998)
32. A. Sugita, RIMS Kôkyûroku, 1507, 147 (2006) (in Japanese).
33. S. Goldstein, J.L. Lebowitz1, R. Tumulka, N. Zanghi. Phys. Rev. Lett. **96**, 050403 (2006)
34. S. Popescu, A.J. Short, A. Winter, Nat. Phys. **2**, 754 (2006)

35. P. Reimann, Phys. Rev. Lett. **99**, 160404 (2007)
36. M. Rigol, V. Dunjko, M. Olshanii, Nature **452**, 854 (2008)
37. A. Lenard, J. Stat. Phys. **19**, 575 (1978)
38. W. Pusz, S.L. Woronowic, Commun. Math. Phys. **58**, 273 (1978)
39. H. Tasaki, arXiv:cond-mat/0009206 (2000).
40. M. Campisi, Stud. Hist. Phil. Mod. Phys. **39**, 181 (2008)

MIX
Papier aus verantwortungsvollen Quellen
Paper from responsible sources
FSC® C105338

If you have any concerns about our products,
you can contact us on
ProductSafety@springernature.com

In case Publisher is established outside the EU,
the EU authorized representative is:
Springer Nature Customer Service Center GmbH
Europaplatz 3, 69115 Heidelberg, Germany

Printed by Libri Plureos GmbH
in Hamburg, Germany